PRAISE FOR *ON THE BINDING BIASES OF TIME*

A collection of essays that reads like a picaresque novel, *On the Binding Biases of Time* takes the reader on a journey into the heart of ecological thinking. Lance Strate—an artist at the process of abstracting— delivers on his promise in the Introduction: for people unfamiliar with the fields he deals with, he provides a very good summary and explanation. But he does much more: for readers well versed in these fields, he provides a GPS—not a map but an entire navigational system—connecting between general semantics and media ecology; between Korzybski, Johnson, McLuhan and Postman; the Ten Commandments and Tolkien; Groucho Marx, Goldilocks and Pete Seger; Heraclitus and postmodernism; World War I and 9/11; consciousness and the self; space and time. And he does this in his usual lucid prose, with a deeply touching poetic streak and wonderful sense of humor. If Neil Postman whom he quotes was right and "clarity is courage," Lance Strate gets the Medal of Honor.
—**Dr. Eva Berger, Dean of the School of Media Studies, College of Management and Academic Studies, Israel**

What a wonderfully compelling and utterly inviting entry point to one of the most significant conceptual frameworks of modern times. Lance Strate should be applauded for bringing Alfred Korzybski and general semantics into the contemporary conversation as never before.
—**Douglas Rushkoff, author of Program or Be Programmed, and Life Inc.**

The Binding Biases Time is a very humane book authored by a thoughtful and insightful writer. Lance Strate brings to life the central concepts of general semantics, systems theory, and media ecology in brief, sometimes poignant, sometimes hilarious essays. This volume provides a better, more engaging discussion of these complex topics than anything else published in the past quarter century. It is packed with important insights about communication, media, and living in today's world. It also happens to be great fun to read.
—**Michael Cole, Dr. Sanford I. Berman Chair in General Semantics at the University of California, San Diego**

Aristotle was wrong. A thing CAN be both A and not-A—and this collection of essays, combining scholarly rigor and compulsive readability, is proof positive. Highly recommended for discerning meaning-makers of every stripe: from the specialist to the student to the simply curious.
—**Susan Maushart, author of The Winter of Our Disconnect**

Anyone who has had the pleasure of hearing Lance Strate deliver one of his keenly illuminating addresses will know what to expect in this volume: daringly original extrapolations of the work of McLuhan, Korzybski, Tolkien (that's right), and a flurry of thinkers familiar and perhaps not yet familiar to you. If you're in search of some intellectual stimulation, coming upon this book is your lucky day.
—**Paul Levinson, author of New New Media, The Plot to Save Socrates, and Twice Upon a Rhyme**

This intelligent and well-crafted collection of essays by Professor Strate provides an essential, complete and accessible context for understanding the contemporary intersection of general semantics, media ecology and the broad array of disciplines that fall under the umbrella of communication studies. *On the Binding Biases of Time* is a "must read" not just for those interested in the historical and conceptual evolution of these fields of study, but for all who want to understand how and why these disciplines are enjoying a theoretical and practical resurgence in importance and popularity for scholars and the general public nearly 90 years after some of these ideas were first introduced.
—**Ed Tywoniak, Professor of Communication, Saint Mary's College of California**

ON THE BINDING BIASES

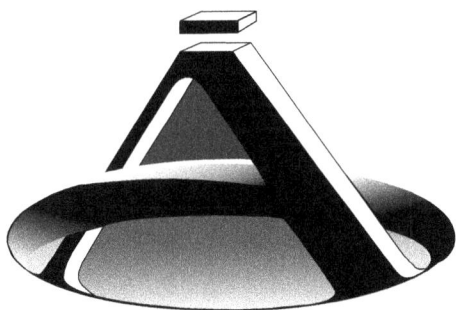

OF TIME

AND OTHER ESSAYS ON
GENERAL SEMANTICS
AND MEDIA
ECOLOGY

Lance Strate

The New Non-Aristotelian Library
Institute of General Semantics

Fort Worth, Texas

Institute of General Semantics
3000 A Landers Street, Fort Worth, Texas 76107
http://www.generalsemantics.org

Book design: Peter Darnell/Visible Works Design

Library of Congress Cataloging-in-Publication Data

Strate, Lance.
  On the binding biases of time and other essays on general semantics and media ecology / Lance Strate.
       p. cm.
  Includes bibliographical references and index.
  Summary: "Explorations of how we use symbols and media to relate to our environment, and how different modes of perception and communication influence human consciousness, culture, and social organization. These essays draw upon and integrate the perspectives of general semantics, systems theory, and media ecology"--Provided by publisher.
  ISBN 978-0-9827559-2-1 (cloth hardcover : alk. paper) -- ISBN 978-0-9827559-3-8 (pbk. : alk. paper))
  1. General semantics. 2. System analysis. 3. Communication--Moral and ethical aspects. I. Title.
  B820.S77 2011
  149'.94--dc22
                          2010054017

The New Non-Aristotelian Library

For Christine Nystrom

TABLE OF CONTENTS

Introduction 3

Chapter One   Alfred Korzybski and General Semantics............ 15

Chapter Two   Quandaries, Quarrels, Quagmires, and
                   Questions..39

Chapter Three   A Systems View of Semantic Environments
                   and Media Environments 53

Chapter Four   On the Binding Biases of Time........................67

Chapter Five   Post(modern)man 99

Chapter Six   Defender of the Word 111

Chapter Seven   Paradox Lost 123

Chapter Eight   The Ten Commandments and the
                   Semantic Environment 143

Chapter Nine   Tolkiens of My Affection............................. 157

Chapter Ten   Poetry Ring 175

Chapter Eleven   Be(a)Very Afraid....................................... 191

Chapter Twelve   The Supreme Identification
                   of Corporations and Persons 201

Chapter Thirteen   Healthy Media Choices 209

Chapter Fourteen   The Future of Consciousness 227

References   251

Index   275

Introduction

The fugitive essays that I have collected in this volume trace varying paths through a common landscape. That territory is navigated first and foremost with the aid of the highly useful map known as general semantics, but also by means of triangulation. That is to say, other maps are also referenced, notably systems theory, and media ecology, both of which are, in part, elaborations on Korzybski's original non-Aristotelian approach. Readers unfamiliar with general semantics (and/or systems theory and/or media ecology) will find sufficient summary and explanation of terms and concepts here so that they need not fear wandering aimlessly about in unfamiliar terrain. My aim in these essays is to communicate in a manner accessible to scholars, students, and an intellectually-minded general readership. I should note, however, that this collection is not meant to be a primer on or a comprehensive survey of general semantics, or systems theory or media ecology for that matter, although the reader that *is* introduced to one or another of these topics in this book will find appropriate recommendations for further reading.

To anyone who undertakes the journey, it ought to be clear enough that there is but one territory that we are traveling through, however many names we give to it, and however many maps we draw of it. But our capacity to create a multiplicity of words and representations has the effect of fracturing the world in our eyes, creating false divisions, and differences that make no difference, thereby blinding us to underlying unities. And so we refer to different aspects of the same reality through terms such as *language, grammar, structure, relationship, system,* and *medium*. And we point to different facets of the same set of phenomena through terms such as *communication, interaction, information, abstraction, perception, signification, symbolization, technologization,* and *mediation*. In the end, we are concerned with the one and only *environment* that we, as human beings, find ourselves within, the single, indivisible *ecology* that we are a part of, the *common ground* we share with the rest of existence.

Each chapter in this book constitutes an individual excursion, and it is certainly my intention, and hope, that readers find each to be worthwhile on its own merits. The paths traced by these fourteen expeditions intersect on many occasions, a certain amount of crossings and circlings being inevitable when we are tracing one-dimensional pathways in an effort to map a two-dimension terrain. So, as we return to familiar landmarks

On the Binding Biases of Time

in these essays, I trust readers will understand that we are coming upon them from different directions, viewing them from different angles and perspectives, in an effort to deepen our understanding of this topography. Rather than a lone intellectual journey that proceeds from beginning to middle to end, this collection creates a network or matrix of ideas that I hope readers will find rewarding in its own way. As such, it would be possible to read these essays in any order, but the bias of the book as a medium requires a linear arrangement. Therefore, I have endeavored to place these essays in a sequence that I believe to be most helpful in guiding readers through this landscape, as well as establishing a route that is aesthetically pleasing.

These geographical metaphors, and the discussion of order and arrangement, obscure the fact that reading and writing, as well as editing and publishing, are activities that take place over time, and that every journey is a journey through time as much as space. I have chosen "On the Binding Biases of Time" as the title essay of this volume for its poetic value, and because it refers to key concepts in general semantics (time-binding) and media ecology (time bias), but also because time is an underlying theme that runs through these essays. This is not a book about time, or the study of time, but general semantics is in many ways about the addition of the dimension of time to modes of thought that otherwise assume an essentially timeless environment. And media ecology is in many ways about the need to understand human environments, and human beings themselves, in terms of time and sound, against the prevailing bias in western cultures toward space and vision, and this includes the need to place contemporary events in an historical context. Time is truly our invisible environment, and perhaps all the more true for being so, and questions regarding our understanding of time, of our past and our future, permeate this collection. I refer to these essays as fugitive, and as such indicate that they may be of only temporary interest to readers, but I do so with the understanding that all things in this world are temporal, bound to a particular time and place, and subject to inevitable change. So I send this book out to a future I cannot know as I write these words, with the understanding that what I have written, however permanent it may seem to me, is nothing more than a fleeting event, or at best, a series of events over a rather limited period of time.

Each of these essays has been subject to a measure of revision

4

and editing from their original form as I prepared them for publication in this volume. Wherever possible, I have endeavored to eliminate or correct errors, and add references, notes, and explanations, and otherwise improve upon the original. The first chapter, "Alfred Korzybski and General Semantics," provides an introduction to the life and times of the founder of general semantics, and to what has come to be known as *Korzybskian general semantics*. This piece originated as an article entitled "Between Two Worlds: Alfred Korzybski and General Semantics," prepared for Volume 42 (2010) of the prestigious Polish journal *Organon*, published by Instytut Historii Nauki, Polska Akademia Nauk (Institute for the History of Science, Polish Academy of Sciences); and there is a certain harmony in this, as Korzybski was a native of Warsaw (which is where PAN is located) who came to the United States, where he formed his non-Aristotelian system of general semantics, and *Organon* is the name originally given to Aristotle's works on logic, meaning, and semantics. I am grateful to *Organon* editor Robert Zaborowski for permission to include this article in this collection, and to Bruce Kodish for his assistance in preparing the piece.

The second chapter, "Quandaries, Quarrels, Quagmires, and Questions," is based on Wendell Johnson's take on general semantics, with the addition of a media ecology perspective to elaborate on the problem of identity relationships as they have evolved over the history of western culture. This essay originated as a paper presented at the *World in Quandaries* symposium, held at Fordham University, New York City, on September 8, 2006, which I organized. The symposium marked the 60th Anniversary of the publication of Wendell Johnson's *People in Quandaries* (1946), along with the 60th Anniversary of the New York Society for General Semantics, and the 8th Anniversary of the Media Ecology Association, and I would like to thank Allen Flagg, President of NYSGS for making the event possible; the paper was published the following year in Volume 64 of *ETC: A Review of General Semantics*, edited by Nora Miller. The next chapter, "A Systems View of Semantic Environments and Media Environments," introduces systems theory, with special emphasis on Niklas Luhmann's work on social systems, as a basis for integrating the work of Korzybski and Marshall McLuhan; the essay introduces the notion of *mode of abstraction*, as a complement to *level* or *order of abstraction*, to general semantics, and *mediating*, as an alternative to *medium*, to media ecology. The earliest version of this essay took the form of an address I gave at the *Across the Generations: Legacies of Hope and Meaning* Conference sponsored by the Institute of General Semantics, cosponsored by the Media Ecology Association, and held at Fordham University on September 11-13, 2009, which I organized. It was subsequently published in Volume 76 (2009/2010) of the *General Semantics Bulletin*, edited by Brian Cogan. A second version was presented as a paper at the 11th Annual Convention of the Media Ecology Association, cosponsored by the IGS, held at the University of Maine, Orono, on June 10-13, 2010, and subsequently published in Volume 11 (2010) of the *Proceedings of the MediaEcology Association*, edited by Ellen Rose.

Chapter Four contains the title essay, "On the Binding Biases of Time," and, as noted above, places special emphasis on the concept of time-binding, the foundation upon which Korzbyski built his non-Aristotelian system, and on the concept of time bias, a key term for Harold Innis in his media ecological study of the material nature and impact of media and technologies. The essay also considers the topic of time more broadly, and how our views of past, present, and future have changed in relation to changes to our media environment. The piece is based on a keynote address I gave at the 67th Annual Conference of the New York State Communication Association held in Ellenville, New York, on October 23-25, 2009, which was organized by Donna Flayhan. One version of this essay was published as "On the Binding Biases of Time: Korzybski, Innis, and the Future of the Present," in Volume 20 (2010) of the *Electronic Journal of Communication*, edited by Teresa M. Harrison. A second version was prepared for the 2009 *Proceedings of the New York State Communication Association* edited by Roxanne O'Connell, and a third, entitled "On the Binding Biases of Time: An Essay on General Semantics, Media Ecology, and the Past, Present, and Future of the Human Species," was published in Volume 67 of *ETC*, edited by Bill Petkanas.

The next two chapters take as their focus Neil Postman, who was my mentor, colleague, and friend, in addition to being an important bridge between the discipline of general semantics and the field of media ecology. The first, "Post(modern)man," which appears as the fifth chapter in this collection, is also the earliest publication to be included here, having originally been presented as a paper entitled, "Post(modern) man, or Neil Postman as a Postmodernist," at the at the 79th Annual Conference of the Speech Communication Association (now the National Communication Association), held in Miami Beach on November 18-21, 1993. The paper was presented as part of a session I had organized entitled *Communication, Education, and Culture: Perspectives on the Scholarly Activity of Neil Postman*. A second version took the form of an address given at the 39th Media Ecology Conference (the in-house conference of the old Media Ecology graduate program at New York University), held in Saugerties, New York on October 7-9, 1994. And another version was published as an essay in Volume 51 (1994) of *ETC*, edited by Jeremy Klein. This piece, written while Postman was still alive, discusses the similarities between his media ecology perspective and that of postmodernists such as Jean Baudrillard, Jean-François Lyotard, and Fredric Jameson, and situates Postman's work as a defense of verbal communication, both spoken and written, in the face of threats posed by image culture and the bias towards quantification associated with the technological imperative. I am grateful to Neil for the guidance and inspiration that he gave me, and for putting up with my attempts to put his work into perspective. Almost a decade later, after he passed away, I was asked by *ETC* editor Paul Dennithorne Johnston if I would write something for the journal to honor his memory. I did so, and with the exception of quotations, wrote the piece in E-Prime (the general semantics innovation in which the English language writer (or speaker) makes no use whatsoever of the verb *to be* in any form or conjugation; I also utilized the general semantics extensional device of dating in the piece. Harkening back to the earlier article, I gave the piece the title, "Neil Postman, Defender of the Word," providing a biographical sketch that emphasizes Postman's involvement in general semantics and the study of language and human behavior, and was grateful for the opportunity to share a quote from Postman that he once said in conversation with me, but never put into print: "Clarity is courage." The article appeared in Volume 60 (2003) of *ETC*, and I also gave a talk based on the essay at the 90th annual meeting of the National Communication Association, held in Chicago, on November 11-14, 2004. It appears here as Chapter Six under the title, "Defender of the Word."

The seventh chapter, "Paradox Lost," was written as the Introduction to *Paradox Lost: A Cross-Contextual Definition of Levels of Abstraction* by Linda G. Elson (2010). Linda was a doctoral student in New York University's Media Ecology program, working under Neil Postman and Christine Nystrom, when she fell ill and was unable to publish her brilliant study. The work was posthumously edited by Alan Ponikvar, and published by Hampton Press of Cresskill, New Jersey, in the Media Ecology book series for which I have been the supervisory editor, and my introduction provides an overview of that portion of the field of media ecology concerned with language and symbolic form, including general semantics. It therefore works quite well taken out of the context of *Paradox Lost*, and recontextualized in this collection, drawing at it does on research conducted for my own book, *Echoes and Reflections: On Media Ecology as a Field of Study* (which was published by Hampton Press in the media ecology book series in 2006). I am grateful to Alan Ponikvar for permission to include my introduction here, and I also urge you to read Linda Elson's *Paradox Lost* —it is well worth your attention. Chapter Eight, "The Ten Commandments and the Semantic Environment," examines the Decalogue as a deliberate attempt to modify the semantic environment, and media environment, of the ancient Israelites, and thereby alter both consciousness and culture. This essay originated as a paper presented at the *Mind and Consciousness* symposium, held at Fordham University in New York City on October 27, 2007, which I helped to organize, and appeared in print in Volume 74/75 (2007/2008) of the *General Semantics Bulletin*, which I edited, under the title of "The Ten Commandments and the Semantic Environment: Understanding the Decalogue Through General Semantics and Media Ecology."

The ninth chapter, "Tolkiens of My Affection," casts an affectionate eye on the fiction of J.R.R. Tolkien, discussing how his background as a philologist, studying languages and literature (with an awareness of orality and literacy issues) informed his creative writing. Originating as a paper presented at the 61st annual New York State Communication Association conference, held in Kerhonkson, New York on October 24-26, 2003, it was revised for publication in Volume 66 (2009) of *ETC*, edited by Bill Petkanas. Chapter Ten, "Poetry Ring," consists of ten short bits about the relationship between general semantics, with its emphasis on a scientific approach to accuracy and clarity in language, and poetry, which serves (among its many functions) as a means of gaining critical perspective on language. In 2008, I began co-editing a poetry feature for *ETC* with Dale Winslow, and we started each section with a set of brief remarks on the connection between poetry and general semantics, and in some instances, on the theme of the poems being featured in that issue. We produced ten of these features, and I have abstracted out of them fragments that fit into the context of this book, taking them out of chronological order in order to place them in a more logical arrangement. Rather than a cohesive essay, my hope is that the impressionistic nature of this chapter reflects the poetic nature of its topic, and I thank Dale Winslow for permission to use this material here in this manner.

The next two chapters originated as posts on my blog, *Lance Strate's Blog Time Passing* <http://lancestrate.blogspot.com>. Chapter Eleven, "Be(a)Very Afraid," a light piece written about the decision of *Beaver* magazine to change its name to *Canada's History*, was posted on January 24, 2010, and Chapter Twelve, "The Supreme Identification of Corporations and Persons," written about the Supreme Court's rulings granting corporations the legal status of persons, was posted on January 27, 2010. Both posts were revised for publication in Volume 67 (2010) of *ETC*, edited by Bill Petkanas, prior to being modified to take their place in this book. Chapter Thirteen, "Healthy Media Choices," consists of an edited transcript of a radio interview with me conducted by Mary Rothschild, that was recorded on May 27, 2010, at the WFUV studios on Fordham University's Rose Hill campus in the Bronx, and aired on WVEW Brattleboro Community Radio in Vermont on June 8, 2010, as that week's *Healthy Media Choices Hour*, hosted and produced by Mary Rothschild. The program is archived as a podcast on the Healthy Media Choices website <http://www.healthymediachoices.org>, and also on Witness for Childhood <http://witnessforchildhood.wordpress.com>. The interview reviews some basic ideas about general semantics, as well as providing discussion on media literacy and education, and progressive spirituality, and I am grateful to Mary Rothschild for her permission to include it here in this collection. The final chapter, "The Future of Consciousness," is a wide-ranging essay that considers the many different and interrelated meanings of the term *consciousness*, and the nature, origin, history, and continued evolution of human consciousness as the internalization of modes of communication. The piece originated as a presentation given at the *Envisioning the Emerging Future* Colloquium sponsored by the Institute of General Semantics, held at the American Museum of Natural History

On the Binding Biases of Time

in New York City on April 23, 2005 (the topic having been suggested by Allen Flagg who had organized the event), and published in Volume 66 (2009) of *ETC*, edited by Bill Petkanas.

As may be apparent from the above summary, much of the material that makes up this book was either produced or given its penultimate form in recent years, a period of time that coincides with my tenure as Executive Director of the Institute of General Semantics. It has been an honor and a privilege to stand in the shoes of Korzybski, follow in his footsteps, and climb up on his shoulders to take in the breathtaking view. I have found this time to be an enriching and enlightening experience, and I am grateful for the ways in which it has contributed to the development of own thought and understanding, deepening my understanding and appreciation for general semantics, as well as systems theory and media ecology. General semantics has much to offer, in a practical way for individuals and institutions, and theoretically and philosophically for the advancement of knowledge. Indeed, it is truly unfortunate that this field is so often overlooked these days, in the academy, and outside of it. I hope that my meager efforts have contributed in some small way to the Korzybski Revival now underway, to a renewal of interest in his non-Aristotelian system, and to its continued progress and evolution. In appreciation, I have asked that all of the royalties earned for sales of this book be donated to the Institute of General Semantics that Korzybski founded in 1938.

When I was in elementary school, attending P.S. 99 in Kew Gardens, New York, one of the many books I bought at our annual Scholastic book fairs was *What's Behind the Word*, by Sam and Beryl Epstein (1964). That wonderful little book captured my imagination with its story of the evolution of the English language, the nature of words, and the role played by writing, the alphabet, and printing, and started me on the road that led me to this place, and time, and for that I wish to say thank you to the Epsteins. I also want to acknowledge Jack Barwind, who introduced me to general semantics, systems theory, and what I would later come to know as media ecology in the first semester of my freshman year at Cornell University, when I took his Introduction to Communication Theory course; I first read *Science and Sanity* in an upper level theory course that he taught. Several years later, upon entering the Media Ecology doctoral program at New York University, I was delighted to discover Neil Postman's involvement with the International Society for General Semantics (now merged with the IGS), and that the media ecology curriculum included systems theory as well as general semantics. Neil's influence is felt throughout this volume, and I would also like to express my sincerest thanks to Postman's colleagues, Terence P. Moran, a dedicated educator, Henry Perkinson, an outstanding scholar, and especially to Christine Nystrom, my mentor and dear friend. There are several other fellow travelers along this intellectual highway who merit a mention as well, including Paul Levinson, Thom Gencarelli, Janet Sternberg, Martin Friedman, Stacy Zeifman, Eva Berger, Mary Alexander, Robert Blechman, Paul Lippert, Geri Forsberg, Thierry Bardini, Robert Logan, Eric McLuhan, Nora Bateson, Mary Catherine Bateson, Deborah Tannen, and Douglas Rushkoff.

In addition to the individuals already named above, I would like to express my gratitude to Corey Anton, supervisory editor of the New Non-Aristotelian Library series published by the IGS, to Peter Darnell of Visible Works Design, and to my graduate student assistants, Andrea McCrary, Ariella Astion, Blake Seidenshaw, Mary Purcell, and Pamela Miller. I am deeply grateful to Fordham University for providing the intellectual environment and scholarly support that has made this work possible, and especially the Department of Communication and Media Studies, including department head James VanOosting, and also Dean of Faculty John P. Harrington, Provost Stephen Freedman, and President Joseph M. McShane, SJ. I also want to acknowledge Congregation Adas Emuno, my source of spiritual support, where I have had the opportunity to explore ideas about language, media, and communication in a context and environment apart from academia and the IGS. And I would be remiss if I did not express my gratitude for the support shown to me by my family, my mother Betty Strate, my wife Barbara, my children Benjamin and Sarah, and our dog, Dingo. No doubt I have omitted some important names here, and for that I apologize, and ask that you accept my simple thank you.

And how else could I end this introduction, but by saying, *et cetera?*

Chapter One
Alfred Korzybski and General Semantics

Alfred Korzybski's life story is a compelling one. Born in Warsaw on July 3, 1879 into a noble family, Alfred Habdank Skarbek Korzybski was the son of Hrabina Helena Rzewuska and Hrabia Ładysław Korzybski, and was later known in the English-speaking world as Count Korzybski. He attended Politechnika Warszawska, where he earned an undergraduate degree in engineering, took some additional postgraduate courses at the University of Rome, and continued his studies in numerous fields on his own. A polymath and pioneer of interdisciplinary studies, his major intellectual influences include Albert Einstein in science; Cassius J. Keyser and Henri Poincaré in mathematics; William Alanson White, Ivan Pavlov, and Sigmund Freud in psychiatry and psychology; and Bertrand Russell, Alfred North Whitehead, Ludwig Wittgenstein, Josiah Royce, and indeed Aristotle, in philosophy.

As a young man, dismayed at the fact that peasants were forced to remain in an illiterate state by Czarist decree, he built a small school for the local peasantry on the country estate where he had grown up, outside of Warsaw; consequently, he was arrested, sentenced to be imprisoned in Siberia, and saved only by his father's intercession. When he was 35-years-old, with the advent of the First World War, he volunteered to serve in the Russian Army, and while he was not awarded any rank above that of private, he was assigned to a Cavalry Detachment of the General Staff Intelligence Department, and had special privileges and duties representing the Second Army Intelligence Department on the battlefield. Korzybski was wounded in the leg and suffered other permanent injuries during the war, and subsequently was sent by the Russian Military Commission to North America as an Artillery Expert. Spending time in both Canada and the United States, he learned how to speak English (already being fluent in Russian, German, French, and his native Polish). When the Russian Revolution resulted in Russia's withdrawal from the war, he joined the French-Polish Army in 1917; he was appointed Secretary of the French-Polish Military Commission and served as a recruiting officer lecturing to Polish audiences in Canada and the United States, and he worked for the United States government's Fuel Administration lecturing to miners to encourage coal production and promote the sale of Liberty Bonds. Not long after the war ended, he met Mira Edgerly, a well known American portrait painter, and they were married in 1919.

He published his first book, *Manhood of Humanity: The Science and Art of Human Engineering*, in 1921 (in 1950, shortly after his death a second edition of *Manhood of Humanity* was published, in which the subtitle was eliminated). After an extended period of writing and research which included studying psychiatric methods firsthand at St. Elizabeths Hospital in Washington, D.C., he produced his magnum opus, *Science and Sanity: An Introduction to Non-Aristotelian Systems and General Semantics* (which some have referred to as Korzybski's *Organon*), which was published in 1933 (and subsequently in four more editions, the last two posthumously, the most recent in 1993). In 1937, a transcript of his notes from lectures in general semantics given at Olivet College in Michigan was published under the title of *General Semantics Seminar 1937* (with two posthumous editions following, the most recent in 2002). Lecturing and conducting seminars without an academic position, he gained a loyal following, and with the support of his students, was able to set up the Institute of General Semantics in 1938 in Chicago.

In the years that followed, he offered regular seminars, lectures, and also personal counseling for his students. In 1940, he became a citizen of the United States, and in 1942, a group of his students founded the International Society for General Semantics (which merged with the IGS in 2004), and began publishing the journal, *ETC: A Review of General Semantics* the following year. There was a great deal of public interest in Korzybski and general semantics at that time, no doubt fueled in part by the Great Depression and the Second World War, and a number of popularizations of his work did much to spread the word about his ideas, which enjoyed great influence in many different areas of study and practice. Following the war, Korzybksi was forced to relocate the Institute to Lakeville, Connecticut, and finances remained a concern for the Institute and for him personally. In 1948, he published an abridged version of his main work under the title of *Selections from Science and Sanity* (posthumously modified over eight printings, and transfer to a CD-ROM, and now in a second edition published in 2010). *Time* magazine printed a brief profile of him in 1949, which began: "Yale University had never had a guest lecturer quite like the count. He was an egg-bald old (69) gentleman who dressed in Army-style sun-tans, refused to wear a coat or tie, and spent most of his time in a wheel chair" (Always, p. 68). (Somewhat flippant in tone, the article was incorrect on the point of the wheelchair, as he typically was able to walk with the help of a cane, and the wheelchair was used only to help Korzybski get from the Yale residence hall where he was staying to the lecture hall in timely fashion, as no cars were allowed on campus at that time.) Further on, the article relates

> Twenty years ago, few would have walked across the street to listen to Alfred Korzybski. Today, as founder of a whole new system of thought called general semantics, he has hundreds of followers all over the world, and the respect of many scientists and scholars. Disciples have written articles on his subject ranging from "General Semantics and Dentistry" to "General Semantics and the Teaching of Physics." Doctors, using general semantics, have claimed it helped cure everything from alcoholism to frigidity. There are General Semantics Societies in twelve cities from Winnipeg to Sydney. Sample members: Architect Frank Lloyd Wright, President George Stoddard of the University of Illinois, Actor Fredric March.
>
> The father of general semantics was born in 1879—the same year, he likes to point out, as Einstein and Stalin. (p. 68)

Underlying this comment is the fact that Stalin would represent to Korzybski (and to most of us) the worst side of humanity, and Einstein (who Korzybski idolized) would symbolize the best that human beings could aspire to.

Alfred Korzybski passed away on March 1, 1950, due to the effects of a mesenteric (abdominal artery) thrombosis he experienced while working at his desk the day before. At that time, the IGS was just readying for publication the first issue of its own journal, the *General Semantics Bulletin*. The Institute established an annual Alfred Korzybski Memorial Lecture, beginning in 1952, which continues on to this day (along with both the *Bulletin* and *ETC*). Interest in Korzybski and general semantics continued to grow during the 1950s, but went into decline during the late 20th century. In 1990, through the efforts of two of Korzybski's closest associates, M. Kendig and Charlotte Schuchardt Read, Korzybski's various articles and papers were compiled and published under the title of *Collected Writings, 1920-1950*, but without much in the way of fanfare or exposure. The 21st century is likely to witness a reversal of fortune, however, as the first major biography of Alfred Korzybski, written by former IGS Trustee Bruce I. Kodish, is currently nearing completion, and signs of a nascent Korzybski Revival have been seen in recent years.

Time-Binding

Like so many of his generation, Korzybski was horrified at the senseless waste and utter carnage of the First World War, the first war to use weapons of mass destruction. And like so many of his generation, he was determined to do whatever he could to prevent such a recurrence of this sort of insanity, and inhumanity. It was this moral imperative that motivated Korzybski, as he reasoned that the prevention of inhumanity would have to be based on a firm understanding of humanity, of what is it that makes us human, and how human beings can achieve their full potential. As an engineer at a time when positivism held sway, and the paradigm shift brought on by Einstein's revolutionary theories about the physical universe was still underway, Korzybski naturally turned to his background in science as he searched for answers. The nineteenth century had seen the introduction of the Laws of Thermodynamics, and with them the idea that energy, as opposed to matter, is the fundamental substance of the universe, and Einstein's famous equation, $E=MC^2$ cemented the view that matter was just a relatively stable form of energy. An engineer, Korzybski also looked at energy as exactly what is needed to get work done, understanding that different types of energy made a difference in the kind of work that could be performed.

From physics Korzybski turned to biology, and reasoned that forms of life could be classified based on how they obtain energy to do their work (their work being mainly the business of survival, growth, and reproduction). The sun serves as the primary source of energy for life on earth, and plants have developed the most sophisticated method of capturing and storing that energy, photosynthesis; this understanding became the basis of the first major class of life that he identified, which he termed *chemistry-binding*. As the plant kingdom is typically contrasted with the animal kingdom, animals constituted his second major class of life, and he noted that animals are distinguished by their mobility, which they use to roam their environment and consume plants, thereby gaining the energy that plant life had stored via photosynthesis. He therefore referred to the second major class of life as *space-binding*. Korzybski's third major class of life is the human species. Although akin to the animal kingdom, and therefore incorporating space-binding, he argued that human beings are also capable of capturing and storing what could be considered another kind of energy: Knowledge. Animals are, for the most part, unable to pass on what they have learned to their progeny, whereas human beings have the ability to accumulate learning from one generation to the next, build on what previous generations have accomplished, and thereby accomplish more, get more work done in a more effective and efficient manner than was previously possible; in short, we can make progress, and Korzybski was working at a time when progress was seen as an unmitigated good. For this reason, he designated human beings the *time-binding* class of life.

Insofar as Aristotle was the first to create a system of biological classification, it is only fitting that this schema represents Korzybski's first step in developing a non-Aristotelian system. Clearly, however, Korzybski was not attempting to create a classification system on the order of Aristotle, or Linnaeus, although he certainly incorporates a Darwinian understanding of evolution. Rather, his interest is in developing a heuristic means for understanding what is distinctive about the human race. In his emphasis on energy, we can see an implicit response to Marx's materialist, scientific (at least according to Marx and his followers) approach, with its emphasis on means of production as the basis of human society. And while remaining unsympathetic to Communism as a political movement, Korzybski did engage in his own critique of capitalism and commercialism, in *Science and Sanity*, and especially in *Manhood of Humanity*, where he presented his most detailed discussion of his three classes of life; there he argues that most of the wealth in the world was produced through the efforts of previous generations, and therefore ought to be accounted as a legacy belonging to all of humanity. It follows that he considered economic inequality to be largely irrational and immoral, and the same holds true for social and political inequalities. But Korzybski did not predict or advocate revolution, but rather called for education. Ultimately, it was Einstein, not Marx, that he looked to for inspiration on how to reform society, and following Einstein's general theory of relativity, he referred to his own work as a *general theory of time-binding*.

Human Engineering

In identifying time-binding as the distinguishing characteristic of the human race, Korzybski also distinguished between different forms of time-binding within our species. In particular, he noted that the process of time-binding was for the most part slow and quite gradual until the introduction of science and scientific method, at which point it became possible to make progress at a geometric rather than arithmetic rate. That accelerated progress was limited, however, to the sectors of society concerned with science and technology, and did not apply to politics, economics, or our social, psychological, and ethical affairs. The criticism that our wisdom has not been keeping pace with our scientific and technological progress was a common one in Korzybski's time (and remains so today), and his solution seems logical enough for his times, that we need to generalize from our successes in science and engineering to society in its entirety, that we need to apply a scientific approach to all facets of human life. It follows that one specific solution to the world's ills that he put forth in *Manhood of Humanity* is to put scientists in charge, to institute government by technocracy. Korzybski was far from alone in taking this position, and there is ample precedent, going all the way back to Plato's *Republic*, with its call for philosopher-kings. In Korzybski's utopia, scientists and engineers would become the new aristocracy, but of course this would be a meritocracy, and in this way humanity will have progressed from an infantile state into one of mature adulthood.

Korzybski opens *Manhood of Humanity* (1950) with the following statement:

> It is the aim of this little book to point the way to a new science and art—the science and art of Human Engineering. By Human Engineering I mean the science and art of directing the energies and capacities of human beings to the advancement of human weal. It need not be argued in these times that the establishment of such a science—the science of human welfare— is an undertaking of immeasurable importance. No one can fail to see that its importance is supreme. (p. 1)

A little later on he adds

For engineering, rightly understood, is the coordinated sum-total of human knowledge gathered through the ages, with mathematics as its chief instrument and guide. Human Engineering will embody the theory and practice—the science and art—of all engineering branches united by a common aim— the understanding and welfare of mankind. (pp. 7-8)

Korzybski goes on to argue for a kind of scientific *noblesse oblige*:

The scientists, all of them, have their duties no doubt, but they do not fully use their education if they do not try to broaden their sense of responsibility toward all mankind instead of closing themselves up in a narrow specialization where they find their pleasure. Neither engineers nor other scientific men have any right to prefer their own personal peace to the happiness of mankind; their place and their duty are in the front line of struggling humanity, not in the unperturbed ranks of those who keep themselves aloof from life. . . . The task of engineering science is not only to know but to know how. Most of the scientists and engineers do not yet realize that their united judgment would be invincible; no system or class would care to disregard it. Their knowledge is the very force which makes the life of humanity pulsate. If the scientists and the engineers have had no common base upon which to unite, a common base must be provided. Today the pressure of life is such that we cannot go forward without their coordinating guidance. But first there must be the desire to act. One aim of this book is to furnish the required stimulus by showing that Human Engineering will rescue us from the tangle of private opinions and enable us to deal with all the problems of life and human society upon a scientific basis.

If those who know why and how neglect to act, those who do not know will act, and the world will continue to flounder. The whole history of mankind and especially the present plight of the world show only too sadly how dangerous and expensive it is to have the world governed by those who do not know.

In paying the price of this war, we have been made to realize that even the private individual can not afford to live wrapped up in his own life and not take his part in public affairs. He must acquire the habit of taking his share of public responsibility. This signifies that a very great deal of very simple work, all pointing in the direction of a greater work, must be done in the way of educating, not engineers and scientific men only, but the general public to cooperate in establishing the practice of Human Engineering in all the affairs of human society and life. (pp. 10-12)

Here at last we see that Korzybski's ultimate aim is not a scientific dictatorship, but rather a democratizing of science, as well as a critique of the science of his day. In this sense human engineering, as an engineering of humanity, is essentially an educational initiative: When everyone is able to think and operate like scientists, humanity will be able to function in a mature, rational, peaceful, and fair manner (and this extends to scientists themselves, who often compartmentalize their work from the rest of their activities, and do not necessarily think and operate as scientists even when engaged in science). The task that Korzybski then set for himself was to develop a pragmatic system for teaching people the practice of scientific method as a form of rational and empirical evaluation of the environment, not as a specialized science, but as a generalized approach that anyone could employ. That system was general semantics, and it was predicated on the understanding that it is our species' unique capacity for language and symbolic communication that makes time-binding possible.

A Non-Aristotelian System

In *Science and Sanity*, Korzybski introduced general semantics as a non-Aristotelian system, with the understanding that it should not be considered the only non-Aristotelian system, that others might come after it (or be identified later). He saw general semantics as complementing the non-Newtonian physics and the non-Euclidean geometry associated with Einstein, and the non-Aristotelian designation is specifically directed towards Aristotle's logic, and more generally the essentialism associated with his philosophy, but not necessarily to Aristotle's entire body of work. And as with non-Newtonian physics and non-Euclidean geometry, Korzybski's non-Aristotelian system was not an outright rejection of Aristotelian logic, which Korzybski believed to be quite useful when employed in its proper context. Rather, he argued that the contexts in which Aristotelian logic applied were rather limited, and that a new mode of thought was needed for the much larger set of contexts in which that logic was not helpful (just as Newtonian physics is useful on earth, but inadequate when dealing with the universe in its entirety). Korzybski believed that most people employed a form of Aristotelian logic in their everyday affairs, not in a formal way as in employing syllogisms, but intuitively, as a matter of their common sense descriptions of the world. Aristotelian logic is a deductive system, and in the same way, most people employ what Korzybski termed an intensional orientation, holding their presuppositions and prejudices as axiomatic, judging what they encounter according to their pre-existing beliefs and opinions, and generally not being open to changing their ideas about the world. Modern science, on the other hand, favors induction in conjunction with the empirical method, and Korzybski's non-Aristotelian system incorporates what he termed an extensional orientation, one that requires suspension of judgment, objective gathering and analysis of facts, and continual reality-testing.

The three classic laws of Aristotelian logic are the Law of Identity, the Law of Non-Contradiction, and the Law of the Excluded Middle. Together, they can be interpreted as representing *a priori* essentialist assumptions about the world, assumptions that people generally take for granted, but that contemporary science does not uphold. These assumptions include the notion that a thing is what it is, permanent and unchanging, and that it is always and everywhere the same, that things can always be evaluated in terms of binary oppositions, either/or categories, so that there are no ambiguities, no grey areas or middle ground, that relationships are static and things are discrete and contained, rather than part of a process. Korzybski's non-Aristotelian system, in contrast, rests on the notion that there are no identity relationships in nature, i.e., that no two phenomena are ever exactly alike, and for that matter, no single phenomena is every identical to itself, insofar as everything is a form of energy and in constant dynamic flux, at least on the subatomic level. Just as Korzybski argued in *Manhood of Humanity* that we need to take into account the dimension of time in considering the debt we owe to the past, as well as our responsibility to the future, he demonstrates in *Science and Sanity* that Aristotle's logic, when interpreted as faithfully representing the structure of the world, implies a frozen reality, ignoring the dimension of time and the change it represents (much as Zeno's Paradox also is based on a misunderstanding of time).

To supersede the laws of Aristotelian logic (except for the instances when they could be appropriately applied), Korzybski put forth three Non-Aristotelian Principles of Thought. First and foremost is the Principle of Non-Identity, which extends both to the world where all phenomena are unique events unfolding in time, and also to the ways in which we perceive, understand, and communicate about the world. It is here that Korzybski's famous phrase, *the map is not the territory*, comes into play. The worst case of mistaken identity that we can be a party to is one in which the symbol is mistaken for its referent, for whatever portion of reality it represents. The metaphor of the map was central for Korzybski, because he understood that, as meaning-makers dependent on language and symbolic communication for our time-binding capability, we are map-makers, making mental maps of the world that we encounter, and sharing them with others. And our most important maps are not drawn, but spoken, and written down, hence Korzybski also insisted that *the word is not the thing it represents*. Or as he was fond of saying, *whatever you say a thing is, it is not*. His Second Non-Aristotelian Principle is the Principle of Non-Allness, which reminds us that our perception and knowledge about any given event or object is necessarily incomplete, and all the more so our depictions and descriptions. Returning to the map metaphor, the Principle of Non-Allness reminds us that maps do not represent all of a territory, and by the same token, words do not say all there is to say about any phenomena, there is always more to be said, and there are some things that words cannot describe. Korzybski's Third Non-Aristotelian Principle is the Principle of Self-Reflexiveness. Whereas reality refers to nothing apart from itself (unless we confer additional meaning onto it), representations have the potential to be self-referential, that is to refer back to themselves or to other representations. So, for example, if we are standing within a territory and looking at an ideal map of that territory, it would contain within it a representation of itself, a map of the map. Ideally, the map within the map would also contain a representation of itself, a map of a map of a map, and so on *ad infinitum*. In the same way, some of our statements may be about the world as we experience it, but we can also make statements about statements, and statements about those statements, and so on. We can react to our reactions, evaluate our evaluations, question our questions, and so forth. This of course may lead us further and further away from reality (or at least our experience of it), and therefore is an inherent characteristic and potential problem of language and symbolic communication. It is also the basis of many a paradox, and Korzybski was conversant with Whitehead and Russell's Theory of Logical Types; Gödel's Incompleteness Theorem too is relevant in this context. Moreover, self-reflexiveness suggests subjectivity, as that ideal map would also include the individual looking at the map, and perhaps even the mapmaker if he or she were present. Korzybski was familiar with Pavlov, who defined a symbol as a sign of a sign, that is, a self-reflexive sign, and along similar lines, self-reflexiveness has been viewed as the basis of what is distinctive about human consciousness (e.g., consciousness of consciousness, or self-consciousness).

Consciousness of Abstracting

In *Science and Sanity*, Korzybski (1993) puts forth *abstracting* as a key term, using the verb form to represent it as a process we engage in, rather than *abstraction* as a thing. Abstracting is a scientific term, in chemistry referring to the extraction of one substance out of another, and more generally to abstract means to remove, to take away. Korzybski used the term in a special, technical sense in his system to refer to a process common to both perception and symbolic communication, sensation and signification, distinguishing between different orders, or levels of abstraction:

> As abstracting in many orders seems to be a general process found in all forms of life, but particularly in humans, it is of importance to be clear on this subject and to select a language of proper structure. As we know already, we use *one* term, say 'apple', for at least *four* entirely different entities; namely, (1) the event, or scientific object, or the sub-microscopic physico-chemical processes, (2) the ordinary object manufactured from the event by our lower nervous centres, (3) the psychological picture probably manufactured by the higher centres, and (4) the verbal definition of the term. If we use a language of adjectives and subject-predicate forms pertaining

to 'sense' impressions, we are using a language which deals with entities *inside our skin* and characteristics entirely non-existent in the outside world. Thus the events outside our skin are neither cold nor warm, green nor red, sweet nor bitter [etc]., but these characteristics are manufactured by our nervous system inside our skins, as responses only to different energy manifestations, physico-chemical processes, [etc]. When we use such terms, we are dealing with characteristics which are absent in the external world, and build up an anthropomorphic and delusional world non-similar in structure to the world around us. Not so if we use a language of order, relations, or structure, which can be applied to sub-microscopic events, to objective levels, to semantic levels, and which can also be expressed in words. In using such language, we deal with characteristics found or discovered on all levels which give us *structural* data uniquely important for knowledge. The ordering on semantic levels in the meantime abolishes identification. It is of extreme importance to realize that the relational [etc]., attitude is optional and can be applied everywhere and always, once the above- mentioned benefits are realized. Thus, any object can be considered as a set of relations of its parts [etc]., any 'sense' perception may be considered as a response to a stimulus [etc]., which again introduces relations, [etc]. As relations are found in the scientific sub-microscopic world, the objective world, and also in the psycho-logical and verbal worlds, it is beneficial to use such a language because it is *similar in structure* to the external world and our nervous system; and it is applicable to all levels. The use of such a language leads to the discovery of invariant relations usually called 'laws of nature', gives us structural data which make the only possible content of 'knowledge', and eliminates also anthropomorphic, primitive, and delusional speculations, identifications, and harmful semantic reactions. (pp. 384-385; words in brackets replace abbreviations that were used in the original text)

Simply put, Korzybski's orders of abstraction begin with the Event Level, which represents whatever is going on (some of Korzybski's followers adopted the acronym WIGO, for What Is Going On), the

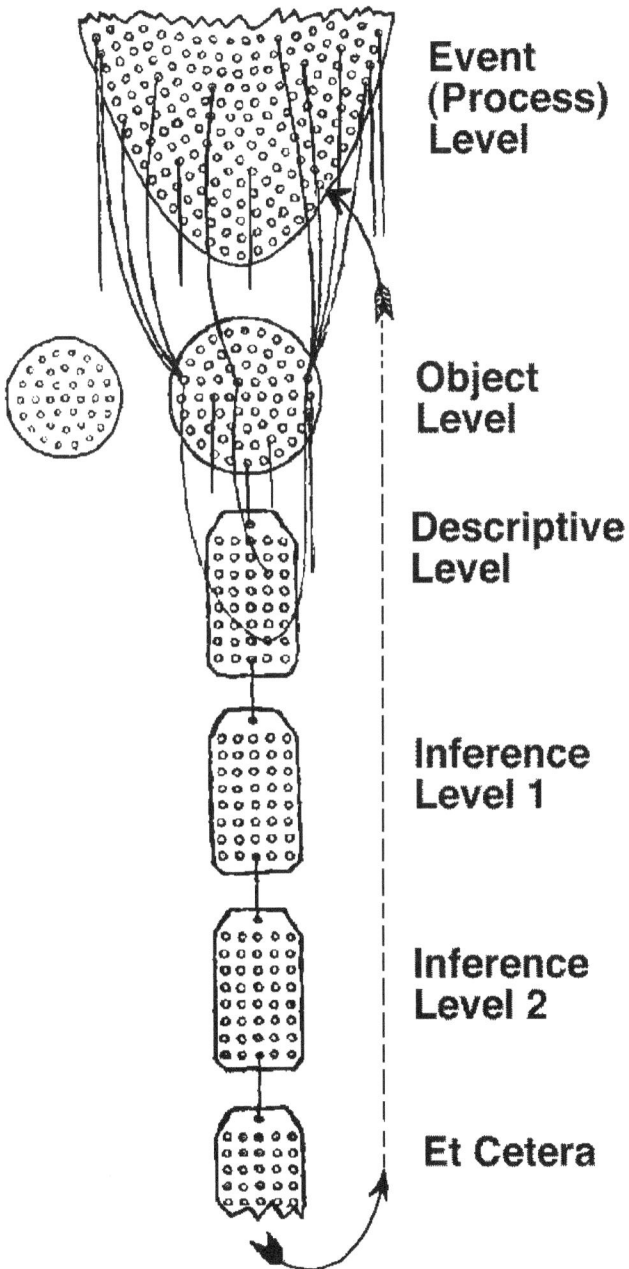

Event
(Process)
Level

Object
Level

Descriptive
Level

Inference
Level 1

Inference
Level 2

Et Cetera

Figure 1. The Structural Differential

reality that we can never quite know. It is only through an excitation or irritation of our nervous systems by stimuli that we gain any indication of the world (and that includes the internal environment of the biological body), and based on such stimuli we build up an inner representation of outer events that may be more or less structurally similar to each other. This process of sense perception constitutes what Korzybski called the Object Level, and at this point, the process is generally the same for animals and human beings. Human beings are unique, however, in that we can continue to abstract, moving from the non-verbal (by which Korzybski meant non-verbal understanding, rather than the more recent concept of nonverbal communication) to Verbal Levels. The first Verbal Level relates to the process of naming and describing, from which we can move on to higher and higher levels of abstraction as we employ self-reflexiveness, or simply assign labels and categories, make inferences and utilize generalizations, and leave out more and more details about the phenomenon in question, a process that invariably involves selection, and therefore subjectivity. Moving up through the levels may also involve drawing conclusions, arriving at opinions, making judgments, and forming beliefs (and ultimately may lead to action, which returns us to the Event Level). And while some words typically function on a higher level of abstraction than others, for example *mammal* is more abstract than *dog*, Korzybski notes that it is also true that the same word may function on different levels of abstraction, and thereby create confusion (for example, *chair* may be used to refer to one specific piece of furniture, or to the entire class of chairs); he termed this quality of verbal communication the *multiordinality* of words.

In order to help individuals to understand the process of abstracting, Korzybski developed a model that he called the Structural Differential. Although it has been criticized as being overly complex, he felt it was an effective teaching tool, especially when used not just as a visual aid, but in its tactile dimension, with fingers tracing the pathways of abstracting (perhaps inspired by the Rosary). In the Structural Differential (see Figure 1), the broken parabola represents the Event Level, of which we only encounter a small part, and the circle stands for the Object Level, the one on the left representing the limited abstracting ability of animals, as it does not continue on into higher levels of abstraction. The tags that follow represent the Verbal Levels. Strings running through the holes stand for the abstracting process, so that as we move through the levels, fewer strings continue. Finally, there is a feedback loop from our high level abstractions back to the Event Level, as the process of abstracting influences the ways in which we act upon the world.

Korzybski's stated goal was *consciousness of abstracting*, that is, he wanted to raise people's consciousness by making individuals conscious of the process of abstracting, and of how Aristotelian thinking takes them further away from any connection with reality, and therefore sanity (which is generally understood as the ability to separate reality from fantasy, and cope with reality). He used the phrase *semantic reaction* to refer to the process by which we assign meanings to phenomena, think and talk about things, and respond to them; in other words, semantic reaction is synonymous with *evaluation*. Through consciousness of abstracting, he wanted individuals to become conscious of their semantic reactions, and to take control of them, rather than to be controlled by them. Following Pavlov, he distinguished between signal reactions, which are reflexive, immediate, kneejerk responses to stimuli, and symbol reactions, which are reflective, delayed, mindful responses. Human beings are unique in our capacity for symbol reactions, which follows from our ability to engage in symbolic communication, and therefore to be fully human, and fully in control of ourselves, we need to learn how to employ symbol reactions, and control our use of signal reactions. In this way, individuals need not act like Pavlov's dogs in response to propaganda, and can avoid psychological and emotional difficulties that stem from inappropriate and dysfunctional personal responses.

Korzybski's use of *semantic reaction*, and *general semantics*, diverged significantly from typical meanings attributed to the term *semantics*, as his concern was not with dictionary definitions, but with meaning, understood as a response to a stimulus, with the way that we come to learn about and relate to our environment. He came to regret his choice of terminology because of the tendency to conflate or confuse *semantics* and *general semantics*, but was never able to come up with a suitable substitute, although he did toy with the terms *general anthropology*, (obviously, this term had already been pre-empted by others), and *up-to-date epistemology* (also *applied epistemology* has been used in reference to general semantics, but of course epistemology is a term that is too obscure for the general audience Korzybski was trying to reach).

Non-Elementalism

Korzybski's non-Aristotelian approach also constituted a shift away from the atomism of the pre-Socratics, and the emphasis on analysis that it engendered, towards a more holistic, ecological perspective, one that emphasized synthesis. As he explains in *Science and Sanity* (1993):

> It is quite natural that with the advance of experimental science some generalizations should appear to be established from the facts at hand. Occasionally, such generalizations, when further analysed, are found to contain serious structural, epistemological and methodological implications and difficulties. In the present work one of these empirical generalizations becomes of unusual importance, so important, indeed, that Part III of this work is devoted to it. Here, however, it is only possible to mention it, and to show some rather unexpected consequences which it entails.
>
> That generalization states: that *any* organism must be treated as-a-whole; in other words, that the organism is not an algebraic sum, a *linear* function of its elements, but always *more* than that. It is seemingly little realized, at present, that this simple and innocent-looking statement involves a full structural revision of our language, because that language, of great pre-scientific antiquity, is *elementalistic*, and so singularly inadequate to express *non-elementalistic* notions. Such a point of view involves profound structural, methodological, and semantic changes, vaguely anticipated, but never formulated in a definite theory. The problems of structure, 'more', and 'non-additivity' are very important and impossible to analyse in the old way. (pp. 64-65)

General semantics, then, is non-elementalistic as well as non-Aristotelian, and in this we can see as well the influence of Einstein's theory of relativity, in which space and time are not absolutes, but rather functions of the relationships among phenomena; moreover, space and time are not actually separable according to Einstein, hence his non-elementalistic neologism, *spacetime*. By the same token, Freud and others argue for the indivisible unity of mind and the emotional and intellectual sides of the mind. Korzybski initially described general semantics as taking an *organism-as-a-whole* approach, and later modified it to further reduce elementalism by identifying the approach as *organism-as-a-whole-in-an-environment*. A non-elementalistic perspective is one that does not engage in oversimplification or overgeneralization, but rather admits to complexity as well as interconnection. As such, it goes hand in hand with what Korzybski termed an ∞-valued orientation, as opposed to the typical one-valued universalism, or two-valued polar oppositions (either/or thinking) of most systems of thought, or even the three-valued orientation of dialectical systems. An infinite- or multi-valued orientation complements the extensionalism of scientific method. And Korzybski's non-elementalism anticipates the development of cybernetics and systems theory (sometimes known as *general system* or *general systems* theory, following the example of *general semantics*). Korzybski, also associates non-elementalism with a non-additive and nonlinear approach, departing even from Newtonian efficient cause (cause-effect); this too is characteristic of systems theory, with its concept of *emergence*, as well as media ecology (e.g., the observations of Marshall McLuhan, 2003, concerning the shift from linear mechanical and typographic technologies to the nonlinearity of the electric circuit and the electronic media).

The Structure of Language

Korzybski (1993) states that "a language, any language, has at its bottom certain metaphysics, which ascribe, consciously or unconsciously, some sort of structure to this world" (p. 89). One aspect of language that he found to be particularly problematic is the verb *to be*. Aristotelian logic begins with the Law of Identity, in which the words *is* and *equals* are interchangeable (*A=A* being the same as *A is A*), and the "*is* of identification" is commonly used in everyday language. So, for example, someone may say that John is a criminal, and that seems to imply that everything there is to say about John is summed up by saying that he is a criminal, and that everything associated with the label of criminal applies to John, that it is a statement structurally identical to saying that one plus one is two. In addition to avoiding the "*is* of identification," Korzybski also advised us to avoid the *"is* of predication," which is associated with adjectives rather than nouns. So, for example, if we tell a child that he *is* bad, we project all of the qualities associated with being bad onto the child, essentially saying that the child is *all* bad and *always* bad. Of course, such practice is frowned upon today, owing in large part to the influence of general semantics. Korzybski was not opposed to the use of the verb *to be* in its entirety, it is important to note, as he saw no problem with its use as an auxiliary verb (e.g., it *is* raining), or simply to indicate existence (I *am*), although some of his followers have attempted to eliminate the use of the verb altogether, at least in written communication. He did want to raise people's consciousness about the metaphysics of *to be*, however, to make them aware of how the verb might lead us astray, and he firmly believed that we would be better off avoiding such Aristotelian usages.

To help us get the non-Aristotelian metaphysics into our nervous systems, he introduced several extensional devices, such as indexing, which involves adding a subscript in order to differentiate between different uses of a word (or levels of abstraction). For example, $chair_1$ is not $chair_2$ is not $chair_3$, as each use of the word *chair* may refer to a specific individual chair, a specific type of chair, the category of chairs in general, or to different meanings (chair as furniture vs. chair as head of a committee, chair as noun vs. chair as verb). Indexing is a way to keep in mind that the same word is being used in different ways, and to keep track of the different meanings. Dating is a similar device, in this case using a superscript to indicate the temporal context of a given phenomenon. For example, Joe^{1980} is not Joe^{2010}, Joe may have been a student then, and a teacher now. $\$20^{2010}$ certainly does not equal $\$20^{1960}$. In addition to contextualization, dating more generally serves to remind us that the world is constantly changing, and that our language does not take that into account.

A third major device used by Korzybski is the *et cetera*, that is, adding *etc.* to the end of a sentence in order to keep in mind the Principle of Non-Allness, that there is always more to be said on the subject, that we cannot help but have left things out of whatever we were saying. Korzybski also thought it important to use hyphens to avoid elementalism, as in the terms *neuro-linguistic*, *neuro-semantic*, and *neuro-epistemologic* that he coined to describe general semantics. And he advocated the use of quotation marks to remind ourselves and others when we are using words on a high level abstraction, underdefined terms, or language that is otherwise ambiguous (e.g., "emotional" maladjustment, "intellectual" achievement). Some have suggested that he invented the nonverbal gesture in which fingers are used to create quotation marks in the air, now in common use, and whether he did or not, he certainly did promote that convention. Korzybski also thought it important to employ qualifying terms when we make statements, for example adding phrases like "it seems to me," "in my opinion," "from my point of view," and also "in our society," "in contemporary culture," etc. In this way, we not only provide further contextualization, but also remind ourselves of the subjectivity inherent in the process of abstracting. On the other hand, employing quantifying terms would lend a measure of objectivity to our statements, for example, stating the exact temperature instead of just saying that it is hot or cold out today. As a measure to be taken against absolutism, Korzybski warned us against allness terms such as *all, always, never, everything, nothing, everyone, nobody, absolutely*, etc., and, in contrast, to employ plurals (e.g., instead of asking *what is the cause of war?*, ask *what are the causes of wars?*).

Following Whitehead and Russell (1925-1927), Korzybksi favored the use of propositions as the only scientifically valid form for language, although he did allow that other uses, such as poetry, could have something to contribute within a general semantics framework. He warned against confusing propositions with inferences, and the general problems caused by the untested assumptions that we make, and about the uncritical use of statements of value, opinion, and judgment, as this type of language could be particularly pernicious. Statements of definition were seen as neutral generally, except when such tautologies are mistaken for propositions. Generally speaking, however, while Korzybski recognized that the capacity for language is the core of our humanity, he expressed a suspicion of words that extended beyond the desire to reform the ways in which we use language, and he was adamant about the need to remain in touch with the non-verbal level of abstracting. And while direct perception was best, the non-verbal visual image was still better than verbal description at representing phenomena, in his estimation (this position is consistent with the fact that modern science is an outgrowth of the visualism of western culture). Ironically, he also argued for the pre-eminence of the highly abstract language of mathematics which, despite its absolute divorce from reality, is entirely unambiguous, propositional, and, according to Korzybski, non-Aristotelian. When applied to phenomena, mathematics allows for unparalleled precision, which is also consistent with a general semantics approach; calculus in particular appealed to Korzybski, in that it was a form of mathematics that tried to capture the dynamic quality of the physical world. Generally speaking, much of the pragmatic success that science lays claim to has involved some form of mathematics, so a privileging of this language is consistent with the idea of general semantics as scientific method writ large.

Korzybski's Ongoing Influence

Korzybski found a sympathetic audience in the United States, where tradition did not have a strong hold on the culture, and great emphasis has been placed on technology, and science. Moreover, his work fit in well with the American philosophical tradition of pragmatism, and Korzybski's was very much an anti-philosophical philosophy, one that was critical of many forms of philosophy as meaningless wordplay. Indeed, his approach was not entirely inconsistent with the American strain of anti-intellectualism, an outgrowth of an anti-elitist democracy, and his aim was to create a school of sorts, a teachable educational system that would benefit everyone, not just the privileged few (and he was able to without being sent to Siberia). Many of his ideas made their way into the American educational system, and one of the great successes of general semantics has been its contributions to education regarding stereotypes, prejudice, and scapegoating, especially during the Civil Rights era.

Korzybski's followers have included a number of prominent individuals, including S. I. Hayakawa, an American (originally Canadian) English professor who went on to become President of San Francisco State University, and then serve in the United States Senate. Hayakawa wrote the most popular book to be published on general semantics, *Language in Thought in Action*, originally published in 1941, and now in its fifth edition (Hayakawa & Hayakawa, 1990). Korzybski was critical of Hayakawa's approach, however, and the two had a falling out. (When asked the reason, Hayakawa answered, "Words.") Another prominent disciple of Korzybski was Stuart Chase, an economist who was highly influential in the New Deal initiative of President Franklin Delano

Roosevelt; Chase published the first popularization of Korzybski's work, *The Tyranny of Words*, in 1938. Another highly significant individual in Korzybski's circle was the prominent speech pathologist Wendell Johnson, who produced an excellent academic introduction to general semantics, *People in Quandaries* (1946); his son Nicholas Johnson was a member of the Federal Communications Commission. Numerous other works summarizing and applying Korzybski's system have been published over the years, including *Drive Yourself Sane*, co-authored by Susan Presby Kodish and Korzybski biographer Bruce I. Kodish (Kodish & Kodish, 2001).

As noted above, Korzybski's work anticipates cybernetics and systems theory, and had a significant impact on Gregory Bateson, Buckminster Fuller, Ludwig von Bertanlaffy, the psychologist Paul Watzlawick, the biologists Humberto Maturana and Francisco Varela, and the sociologist Niklas Luhmann. In the field of computing, Korzybski's ideas helped to shape the thought of Douglas Engelbart and Alan Kay, developers of the Graphical User Interface (GUI)—the Windows and Macintosh interfaces that are ubiquitous today can therefore be traced back to Korzybski's non-Aristotelian system and visualist orientation. In many ways, Korzybksi's general semantics prefigures the development of cognitive science, as can be seen, for example, in the work of Douglas Hofstadter. Korzybski's influence in psychotherapy and the human potential movement was quite significant as well, and extended to Albert Ellis, founder of Rational Emotive therapy; Fritz Perls, founder of Gestalt Therapy; Richard Bandler, founder of Neuro-Linguistic Programming (NLP); and also to Tony Buzan, the British inventor of mind maps; Alan Watts, a British popularizer of eastern mysticism; and L. Ron Hubbard, founder of Dianetics and Scientology. His work has been incorporated into the field of communication, and the study of journalism, and he also has had an impact of the literary world, as William S. Burroughs was one of his students, and Korzybski and general semantics appear in the writings of science fiction novelists A. E. Van Vogt, Robert Heinlein, and Robert Anton Wilson, and less directly in Frank Herbert, Samuel R. Delaney, and Phillip K. Dick; moreover, general semantics is featured prominently in the 1965 French film *Alphaville*, directed by Jean-Luc Godard, as well as Alfred Hitchcock's 1963 Hollywood motion picture, *The Birds*. Korzybski also influenced literary and cultural theory, from

On the Binding Biases of Time

Kenneth Burke's rhetoric, to poststructuralists such as Jacques Derrida and Michel Foucault, and postmodernists such as Jean Baudrillard, Jean-François Lyotard, and Fredric Jameson.[*]

Korzybski's general semantics parallels the linguistic relativism of Edward Sapir, Benjamin Lee Whorf, and Dorothy Lee, the cultural-historical psychology of Lev Vygotsky and Alexander Luria, the semiotics of Charles Saunders Pearce, and C. K. Ogden and I. A. Richards, and the philosophical work of Alfred North Whitehead, Bertrand Russell, and Ludwig Wittgenstein (whom he credits as influences in *Science and Sanity*). Common to all of them is the idea that the structure of our mode of communication has much to do with our thought and behavior, individually and collectively, and this is the basis of the field that has come to be known as media ecology. Within this field, Korzybski is acknowledged by Lewis Mumford and Marshall McLuhan, and appears prominently in the writings of Neil Postman, who served as editor of *ETC* for a decade (see, for example, Postman, 1976, 1988, 1995; Postman & Weingartner, 1969). As originally identified by Postman, Korzybski is considered to be part of the media ecology intellectual tradition, as well as the founder of the discipline of general semantics (Strate, 2006).

General semantics is consistent with the fallibilism of Charles Saunders Pearce, John Dewey, and Karl Popper, but has also been likened to Buddhism and Taoism in its severe critique of language. There is an inherent tension in Korzybski's approach between the emphasis on empiricism extended to everyday life, and the more esoteric exploration of individual consciousness. Between the outer world of the environment and the inner world of the mind rest our forms of mediation, our senses, images, and words, and it is along that boundary, and bridge, that Korzybski labored, in an effort to reconcile the two worlds, and thereby fulfill the promise of humanity's potential.

---

[*] For a fascinating discussion of Korzybski's influence on several aspects of the contemporary intellectual landscape, see Bardini (2009/2010).

Chapter Two

Quandaries, Quarrels, Quagmires, and Questions

N eil Postman, who formally introduced the term "media ecology" in 1968, was known to remark that media ecology is general semantics writ large (Moran, 2007/2008; Postman, 1974). Wendell Johnson's *People in Quandaries* (1946) was required reading in the doctoral program in media ecology that Postman founded at New York University in 1970, no doubt because it provides an accessible and comprehensive introduction to general semantics, and to scientific method. I assume that he did not introduce his students to general semantics by assigning Korzybski's *Science and Sanity* even though it is the original source because he thought that the book was too hard. I also assume that he did not introduce his students to general semantics by assigning Hayakawa's *Language in Thought and Action* even though it is the most popular general semantics work ever written because he thought that the book was too soft. In other words, my Goldilockean conclusion, if you can bear it, is that Postman thought that *People in Quandaries* was just right.

Ecological Thinking and Scientific Method

The story of the Trojan Horse is a well known tale of deception and betrayal, but it is also a classic example of the disastrous consequences of mistaking a symbol for reality. Clever Odysseus, that great manipulator of symbols, knew that the war-weary Trojans would interpret the meaning of the wooden horse *intensionally*, that is, in accordance with their own needs and desires. They would therefore be eager to see the horse as a sign that the Greeks had abandoned their decade-old quest to sack their city, and had set sail for home. The horse was the sacred symbol of the sea-god Poseidon, and Odysseus knew that the Trojans would revere it as a holy icon, and not suspect that it was a false idol. Had the Trojans adopted an *extensional* orientation and engaged in reality-testing, they might have discovered that the Greeks had not sailed across the Mediterranean, but were merely hidden nearby. This in turn might have led them to investigate the horse itself, and determine its true nature as a false front. But after ten years of living with a siege mentality, the last thing the Trojans wanted to do was to look a gift horse in the mouth.

Of course, there were a few Trojans who questioned the symbol of the wooden horse, and the inferences that others had made about its meaning. One of the skeptics was the tragic seer Cassandra, who had the gift of true foresight, but had been cursed so that no one would take her seriously, and most thought her insane. Another was the priest Laocoön, who issued the warning to "beware of Greeks bearing gifts." But Poseidon, who sided with the Greeks, sent serpents to kill him and his sons, and the Trojans took this as a sign that his suspicions concerning the totem were not only incorrect, but also downright blasphemous. And so it came to pass that those who questioned the Trojans' reaction to the symbol, their definition of the situation, and their construction of reality were labeled as being either mad or bad. And, for want of a general semanticist, or media ecologist, the kingdom of Troy was lost.

Over three millennia after the fall of Troy, another set of visionaries warned us to beware of Greeks bearing gifts. Their names were Alfred Korzybski, S.I. Hayakawa, and Wendell Johnson, among others, and the particular Greek that concerned them was not the cunning ruler of Ithaca, Odysseus, but the equally intelligent philosopher from Athens, Aristotle. Aristotle's Trojan horse was symbolic logic, a mode of expression and cognition that misrepresents reality at the same time that it opened the door to most scholarly and scientific investigation. I should note that no one considered Aristotle to be either mad or bad, or an enemy. In fact, Wendell Johnson wrote that if Aristotle were alive today, he would not be an Aristotelian. Instead, he would acknowledge that the time had come to replace his old approach with one that Korzybski had named general semantics; Korzybski characterized general semantics as a non-Aristotelian system, following the example of mathematics, where non-Euclidean geometries had been introduced, and the example of physics, where Einstein's theory of relativity had ushered in a non-Newtonian view of the universe. These three developments are in fact related to one another, and stand in contrast to an older Aristotelian-Euclidean-Newtonian worldview, a worldview in which "things" are solid, discrete, and independent of one another; where reality is static and unchanging; perfect order reigns over chaos and entropy; where species of life are eternal, neither evolving nor becoming extinct; numbers never become irrational, geometries don't go fractal, and mathematical systems do not have to be incomplete if they don't want to be; it was a worldview in which rationality rules the mind rather than unconscious impulse; space, time and truth are absolute, not relative; and meaning and logic are never fuzzy.

In contrast, a non-Aristotelian, non-Euclidean, non-Newtonian worldview is one that emphasizes change and growth, complexity and uncertainty, nonlinear processes and dynamic interactions, interrelationships and interdependence. In other words, it is an ecological worldview.

Korzybski, Hayakawa, and Johnson were engaged in ecological thinking when they explored the relationship between human beings and their symbols, and between symbols and the reality they are thought to represent. They therefore could be placed in the same class as the nineteenth century zoologist Ernst Haeckel, who was concerned with the relationships between organisms and their natural environments, and who coined the term ecology. Another member of this class would be Albert Einstein, whose theory of relativity replaced Newtonian absolutes with a focus on the relationships among physical phenomena. This class would also include the philosopher Martin Buber, who wrote about the relationship between human beings and God, the psychologist Carl Rogers, who emphasized the relationship between therapists and their clients, the educationist Paolo Friere, who argued for the importance of the relationship between teacher and student, and the communication theorist Paul Watzlawick who explained that interaction is more about establishing and maintaining relationships than it is about exchanging content. And this class includes media ecologists such as Marshall McLuhan, Walter Ong, and Neil Postman, as well as others such as Lewis Mumford, Susanne Langer, Harold Innis, and James W. Carey. I have provided an overview of this intellectual tradition in *Echoes and Reflections: On Media Ecology as a Field of Study* (Strate, 2006).

Formal systems of ecological thought, such as media ecology and general semantics, are a relatively recent phenomenon, but ecological thinking itself has been with us throughout our history. Odysseus was an ecological thinker, as was his countryman Heraclitus, a pre-Socratic philosopher who lived not far from where the Trojan War had been fought by his ancestors; his well known statement that you can never step into the same river twice, is quoted with approval by Wendell Johnson in *People in Quandaries* (1946), who writes that "Heraclitus was over two thousand years ahead of his time. The notion which he so aptly expressed has about it a distinctly modern flavor. It is one which Einstein might heartily endorse. It is the basic notion of science, and science as we know it is not as old as Heraclitus—far from it" (p. 23).

On the Binding Biases of Time

What Johnson (1946) meant by "science as we know it" was not so much the science of Copernicus, Galileo, and Newton, but the science of the twentieth century. Johnson describes the modern scientist as "a master of discrimination," explaining that "differences are his stock in trade, and differentiation is the operation by which he performs his wonders" (p. 38). "A similarity," Johnson explains, "is comprised of differences that don't make any difference" and "when a scientist says that two things are similar, he is saying … that certain differences between them do not serve to make them different one from the other, for certain purposes" (p. 38). Similarities, according to Johnson, are never absolute. Consequently theories and generalizations, which are based on perceived similarities, must always be tentative and open to refutation and falsification. Along these lines, Johnson describes the method of science as consisting of:

> (a) asking clear answerable questions in order to direct one's (b) observations, which are made in a calm and unprejudiced manner, and which are then (c) reported as accurately as possible and in such a way as to answer the questions that were asked to begin with, after which (d) any pertinent beliefs or assumptions that were held before the observations were made are revised in the light of the observations made and the answers obtained. Then more questions are asked in accordance with the newly revised notions, further observations are made, new answers are arrived at, beliefs and assumptions are again revised, after which the whole process starts over again. In fact, it never stops. Science as method is continuous. All its conclusions are held subject to the further revision that new observations may require. It is a method of keeping one's information, beliefs, and theories up to date. It is, above all, a method of "changing one's mind"— sufficiently often. (pp. 49-50)

Johnson (1946) goes on to observe that much of what he has described as the method of science has to do with the way that language is used, from which he concludes that "the language of science is the better part of the method of science" (p. 50). He then adds that "the language of sanity is the better part of sanity" (p. 50), by which Johnson means that the goal of general semantics is to adapt the language of science for use in everyday life. To this we might add that the goal of general semantics is to encourage ecological thinking in everyday life. We might further add that the goal is to encourage media ecological thinking, for as Johnson explains about the structure of language:

> On the one hand, it plays a role in determining the structure of our culture, our society, our civilization. On the other hand, it serves as the chief medium or means whereby the individual acquires or *interiorizes* that culture structure. Thus, a study of language structure leads both to a deeper understanding of our civilization and its problems and to a keener insight into the basic designs of individual lives and personalities. It is as though mankind had spun an enormous web of words—and caught itself. (p. 18)

Media ecologists tend to view language as a medium, and often understand media to be technologies and techniques. Consistent with this approach, Johnson (1946) views language as both medium and technique:

> Before we can change our language, it is essential that we develop a certain kind of attitude toward it — the attitude that language is to be viewed as a form of behavior and that, like other behavior, it is to be evaluated as *technique*. . . . we evaluate a technique by asking what it is designed to do, how well it does it, and with what consequences. (p. 269)

Media ecologists also understand media to constitute environments; in one sense they are webs that we create, inhabit, and find ourselves imprisoned by. Accordingly, Johnson (1946) employs the term "semantic environment" (p. 412; see also pp. 417- 426), which we can understand in relation to the larger media environment that includes all of our modes of communication, all of our codes and symbols systems, all of our techniques and technologies. Accordingly, we can could define general semantics as the study of semantic environments, and even refer to general semantics as a semantic ecology.

The Bias of Identity

Johnson, like Korzybski before him, understood that the structure of language as a medium, technique, and environment, is not neutral, but has an inherent bias. Fundamentally, language is a means by which we impose a sense of order, stability, and predictability on an otherwise chaotic, volatile, and uncertain world. It is a method for reducing differences down to a manageable number by directing our attention to similarities. It is a way to gain a sense of control by giving us the power to impose names and labels on phenomena. Language allows us to step into the same river twice, at least symbolically. The bias of language is the bias of identity, and identity is a relationship that exists only in symbols systems. There are no identity relationships in physical, chemical, or biological systems, where no two things or phenomena are ever exactly alike. But language allows us to make identity statements such as one plus one is two, the sky is blue, Pluto is not a planet, war is peace, freedom is not free, ignorance is bliss, and a rose is a rose is a rose.

The bias of identity allows language to function as a kind of informal science, a way of knowing the world, a form of theory-building. And there should be no doubt that the bias of identity has had enormous survival value for our species, serving as a shortcut for making evaluations and predictions about our environment, and helping us to alter our environment to enhance our own survival. The bias of identity is also vital for maintaining social cohesion, inducing cooperation among individuals, and facilitating collective action, without which human survival is impossible; this is why Kenneth Burke (1950), influenced as he was by Korzybski, argues that the primary function of rhetoric is identification, not persuasion. The bias of identity is therefore not a problem in and of itself, and in fact constitutes an evolutionary advantage that has much to do with the success of our species. The problem with identity, I would suggest, is the problem of too much of a good thing. It is the ecological problem of losing a healthy balance. How does this happen? First, we need to recognize that while the bias of identity may be characteristic of language in general, different languages may differ in the degree to which they exhibit this bias. As Johnson (1946) contends, it is possible to reduce the level of this bias in English and other languages. By the same token, the level can be raised, perhaps deliberately by the propaganda techniques that George Orwell described in *1984*, but also accidentally, as the unintended effect of other types of changes. And the most significant change that has affected human language is the invention of writing (Goody 1977, 1986; McLuhan, 1962, 2003; Ong, 1967, 1982).

As a speech pathologist, Wendell Johnson would certainly agree that human language is essentially speech, and he would appreciate the distinction between the spoken word on the one hand, which has been with us for perhaps one hundred thousand years or more, and the written word on the other hand, whose first awkward appearance was only about five thousand years ago. He might even note that the fact that we say that a written word *is* a word, rather than saying that it stands for or represents a word, reflects how deep the bias of identity extends to writing. Writing is a secondary symbol system used to symbolize the primary symbol system of speech. And as a medium, technology, and environment, writing has its own biases, which in turn act on and alter speech and language. One of these effects has been the intensification of the bias of identity. The classicist, Eric Havelock (1978), has demonstrated this change by studying the effects of the alphabet on the ancient Greek language. In the Greek colonies on Asia Minor, the same region where the Trojan War was fought, the alphabet was used to transcribe the oral tradition concerning those events, which we know as the *Iliad* and the *Odyssey*. The content of these poems is essentially preliterate, and as Havelock explains, the language is one of dramatic action, of agents performing acts, rather than statements of static description. The verb "to be" is not used to identify or equate in the language of the oral epics, but begins to be used in this fashion as more and more literate works are produced, that is, its use increases as we move from Homer to Hesiod, through the pre-Socratics, to Plato and Aristotle. Aristotle's logic, which says that, if A equals B and B equals C then A equals C, is in fact a by-product of the ABCs.

The alphabet was first developed by the Semites, and the Greeks learned about this technology from the Phoenicians, which is why they referred to it as Phoenician or phonetic writing. From another group of Semites, the Israelites, came the God of the alphabet, the eternal, all-powerful and unchanging God whose name is represented by four Hebrew letters Yod Hay Vav Hay (YHWH), commonly rendered in English as Jehovah. These four letters are translated as, "I am that I am," a statement of absolute identity that stands as the foundation of monotheism, of the Abrahamic religions of Judaism, Christianity, and Islam. And it was paralleled by the sacred written texts that when copied with care, could be duplicated with little or no variation. Along the same lines, in the Greek colonies on Asia Minor, an oral tradition consisting of countless oral performances over many generations, each one different and unique, was transformed into a fixed text, encoded by means of alphabetic writing; the result was that the extemporaneous and improvisational singing of tales was replaced by a new practice of verbatim memorization and recitation. The variation that was taken for granted as a characteristic of oral tradition has suddenly been thrown into sharp relief by the alphabet, and had come to be seen as corruption, while identity became associated with authenticity (Kirk, 1962).

In the kingdom of Lydia, bordering the Greek colonies on Asia Minor, the alphabet effect led to the minting of the first coins, establishing the idea that all goods can be reduced down to the same monetary units, just as all speech could be reduced down to the same set of twenty-odd letters. Is it any accident that the Greek colonies also gave rise to the first physicists, natural philosophers who introduced the idea that all of the universe could be broken down into identical, indivisible units they called atoms (Logan, 2004). Heraclitus is often counted among them, although he was unique in his emphasis on change and therefore his resistance to the bias of identity. The pre-Socratics laid the groundwork for Aristotle's logic, not to mention Euclid's geometry, while further to the east, the Hindus, who also adopted the alphabet from the Semites, used it to develop the positional numerical notation that we are all familiar with, and with it higher mathematics (Logan, 2004). All of this culminates in Newtonian physics, and the Aristotelian-Euclidean-Newtonian worldview.

We should further acknowledge that the Semites also introduced the concept of law, the earliest examples being associated with the Babylonian Hammurabi, and the Israelite Moses (Logan, 2004). And with formal, written law came the idea that we are all equal and identical before the law. The Greeks, in turn, introduced the concept of democracy, that citizens are the atoms of society, each having an equal say in making political decisions. From these seeds emerge the modern idea of individualism, and with it the declaration that "all men are created equal." The ideal of equality associated with the founding of the American republic and the European Enlightenment presupposes identity relationships among citizens, at least in the symbolic realms of politics and the law, leading to further demands for equality in our social, educational, and economic systems. While modernity was associated with equality through uniformity, contemporary postmodern culture seems to instead favor equality through diversity, the idea that we are all identical in being equally different from one another.

Identity is not just a symbolic affair, as the technologies of mass production have given us a multitude of seemingly identical products. Mechanization begins to take command in the monasteries of medieval Europe, where the invention of the mechanical clock produced the first multiple, identical units, in this case hours, and later minutes and seconds (Mumford, 1934). It continues its march during the fifteenth century in a shop in Mainz, Germany, where Johann Gutenberg starts the printing revolution by producing multiple, seemingly identical copies of the Bible and other texts (Eisenstein, 1979). And it completes its takeover with the Industrial Revolution that begins in the late 18th century and culminates in the early 20th century technique of the assembly line. Mechanization and industrialization also give us the media of mass communication, newspapers and magazines, movies and recordings, and especially broadcasting. These powerful technologies made possible the creation of the mass society, a society in which a mass of individuals are identical in their anonymity and apathy, equal in their alienation and impotence, and all the same in their indifference (Ellul, 1965). This was the moment that Korzybski introduced his non-Aristotelian system, having witnessed the first use of weapons of mass destruction during the First World War. And Wendell Johnson gave us *People in Quandaries* following the even more massive and indiscriminate destruction of the Second World War, in which whole populations were identical in being subjected to concentration camps, gas chambers, aerial bombardments, V-2 rockets, and atomic bombs. It may well be true that every war dating back to the Greek assault on Troy is a war of (or for) identity (McLuhan, 1976), but the two World Wars were wars of mass identity, while the Cold War ended with a massive identity breakdown on the part of the Soviet bloc.

If terrorism and the war on terror represent a different kind of warfare, one fought with new weapons and with the aid of new technologies of communication, they remain conflicts over identity relationships. And our present-day identity politics is just identity war by other means. Along the same lines, electronic technologies have reversed some of the characteristics of mass society, but we have gone from mechanical reproduction to an even more perfect form of digital reproduction, from printing to photocopying to computer-based copy and paste operations.

On the Binding Biases of Time

Our new media continue to extend the bias of identity into new realms. And then there is the biotechnology of cloning, which opens up a new universe of identity relationships. The bias of identity has mutated since the time of Aristotle, but if anything, it has resulted in an identity crisis of unprecedented proportions. And that is why, sixty years later, we still need to read *People in Quandaries*.

The Question is the Solution

In that book, Wendell Johnson wrote about the IFD disease, which stands for Idealization, Frustration, and Demoralization. The IFD disease is a disease of language, brought on by the bias of identity. It begins when we idealize a word, such as love or success, or freedom or democracy. As goals, these vague ideals are unobtainable, no matter how hard we strive for them. And because they are unreachable, we wind up frustrated, and ultimately demoralized. Johnson's solution is to use the language of science, define our terms in a clear, precise, and concrete manner, specify the context in which these terms will be used, and specify the operations and procedures related to these terms. Johnson presented the IFD disease as a quandary, and general semantics as a solution. I want to add a footnote to the IFD disease with the 4 Qs that make up the title of my essay, Quandaries, Quarrels, Quagmires, and Questions (I believe that Johnson, as a specialist in speech, would appreciate the alliteration, as well as my particular emphasis on the peculiar quality of the letter Q). Beginning with Johnson's key term of Quandaries, my intent is to emphasize not just the personal maladjustment that was Johnson's focus, but the interpersonal and social maladjustment that can also occur. To use the example of the Trojan War, which begins when Helen, the wife of Menelaus, runs off with Paris to Troy, the quandary in this case had to do with the idealization of terms such as love, and marriage, both of which remain quandaries in need of operational definitions to this very day. But in this instance, the quandary led to a quarrel, specifically the Greek assault on Troy. The quarrel then resulted in a quagmire, as ten years go by with no resolution to the conflict.

Now, as I mentioned earlier, Odysseus was an ecological thinker. He therefore recognized that the Greek efforts to push through the walls of Troy, coupled with the Trojans resisting by pushing back at the Greeks, had resulted in a stalemate. In effect, the Greeks and Trojans together had created a homeostatic system (Postman, 1976; Watzlawick, Bavelas, & Jackson, 1967; Watzlawick, Weakland, & Fisch, 1974). The Greeks would try to change that system by fighting harder, but this would result in the Trojans fighting back with greater effort, so that the initial change within the system would result in no real change to the system. As an ecological thinker, Odysseus was able to ask the right questions, questions being the fourth Q, and the way out of the quandary. He was able to ask questions about why the Greeks' strategy had failed, and what new strategy might succeed. And he was able to ask questions about how changes within the system differ from changes to the system itself, and how changes within the system might fail, and changes to the system might succeed. And so, Odysseus was able to step outside of the system, instructing the Greek forces to appear to fall back instead of continuing to push forward. The result was that the entire system of Greeks and Trojans stuck in a quagmire experienced system-wide change of epic proportions.

Wendell Johnson stressed the importance of asking good questions, and that is why I have highlighted questions as the means by which we may escape our quandaries, quarrels, and quagmires. In other words, questions are the answer. As Johnson wrote in *People in Quandaries* (1946), "in the meaningful use of language it is a cardinal rule that the *terminology of the question determines the terminology of the answer*" (p. 52). Media ecologists of course recognize that this is another way of saying that the medium is the message. Johnson goes on to explain

> One cannot get a clear answer to a vague question. The language of science is particularly distinguished by the fact that it centers around well-stated questions. If there is one part of a scientific experiment that is more important than any other part, it is the framing of the question that the experiment is to answer. If it is stated vaguely, no experiment can answer it precisely. If the question is stated precisely, the means of answering it are clearly indicated. The specific observations needed, and the conditions under which they are to be made, are implied in the question itself. As someone has very aptly put it, a fool is one who knows all the answers, but none of the questions. (pp. 52-53)

General semantics and media ecology have many good questions,

questions about differences, about what differences make a difference, and what differences may be safe to ignore. Questions about how symbols represent reality, how words stand for and point to things in reality, how maps depict territories, and how media extend us outward into our environments. Questions about what symbols fail to say about reality, what words cannot express about things, what details maps leave out, and how media insulate us from our environment. And questions about the nature of symbols themselves, about what a word is and is not, about how maps are made, about the meaning of meaning and the biases of technologies, about how the medium is the message, and how media, by separating us from our environment, become our new environment. All of these questions are not only good questions, they are ecological questions. They are questions about our relationships with ourselves, with each other, with our symbols and tools, with our semantic environments and media environments. Ultimately, they are questions about achieving sanity on a personal and global level, they are questions about what it means to be human and especially what it means to be human in a technological age, and they are questions about our place in a universe that is 14 billion years old.

Chapter Three

A Systems View of Semantic Environments and Media Environments

A Systems View of Semantic Environments and Media Environments

The aim of this essay is to consider some of the common ground shared by Alfred Korzybski and Marshall McLuhan, and how each scholar's perspective could be used to enhance the approach of the other. In doing so, I rely upon systems theory to bridge the gap between the two scholars, and therefore begin with a brief review of the basics of systems theory, noting how it relates to media ecology and general semantics.

A Systems Overview

The key concept of a *system* is a general one that can be applied to any phenomenon, be it physical, chemical, biological, social, psychological, or technological. A system is any entity that is composed of interdependent parts, so that the whole cannot be explained only by examining the different parts; rather, you have to look at how they work together. In this way, the whole is greater than the sum of its parts, that extra something being the structure, relationship, or interaction among the parts, what Buckminster Fuller termed their *synergy* (see, for example, Fuller & Applewhite, 1975). This something extra that the parts produce when they form a system is said to *emerge* out of the system, and the this quality of *emergence* is held in contrast to linear cause-and-effect because the interaction of the parts of a system, especially one that is complex and dynamic, includes random, chaotic factors that make prediction impossible (although bifurcation points may generate an orderly pattern when repeated over time, just as flipping a coin once generates a random result, but when repeated over and over reveals a statistical order). Changes introduced to a system therefore tend to be difficult to analyze in terms of causal relationships. For example, in some instances a system can take great damage and still go on functioning, as the interdependence of its parts makes for a robust system capable of a high degree of compensation; and in other instances a very small change can bring down the entire system, as interdependence results in a series of indirect effects that may snowball, growing geometrically, especially if the effect forms a feedback loop or is otherwise reiterated, leading to what is sometimes called the *butterfly effect*. Again, the more complex and dynamic a system is, and the more interdependent the parts, the harder it would be to predict the effects of any change that might be introduced into that system. This is particularly relevant for media ecology in regard to the impact of technological innovations.

Systems have boundaries that separate themselves from their environment, and the boundaries can be more or less closed or open to their environment, so that we can distinguish between closed systems and open systems. The universe is considered a closed system, insofar as there is supposedly nothing else outside of the universe (or if there are other universes, nothing outside of the collective multiverse). A biosphere, or a vehicle or suit used in underwater and outer space environments creates a relatively closed system, although it would be open enough to allow in sunlight and other forms of energy. Some social systems are relatively closed, such as a fraternity, military organization, and religious order, where elaborate initiation procedures are required for membership, while others are relatively open, such as a political party, or a social networking site like Facebook or MySpace. Some systems are created by outside forces, for example when we create technological systems, but other systems form spontaneously, "pulling themselves up by their own bootstraps" as the saying goes. Self-organizing systems are referred to as *autopoietic*, the process as *autopoiesis*, and this is associated with the phenomena of chaos and complexity. Such systems emerge out of their environment, which may be another system, so that systems themselves may be the product of *emergence*. Examples of autopoiesis include autocatalytic reactions in chemistry, the (presumed) origin of life as a form of self-organization among protein molecules, and situations where individuals get together to form an organization or society. To create itself, a system needs to construct a boundary to separate itself from its environment, to close itself off from the world in significant ways.

This very brief summary does not do justice to systems theory, but hopefully is sufficient for anyone not familiar with the perspective to have a sense of what I am referencing. Alfred Korzybski (1993) actually anticipates some of systems theory in the 1933 work that introduced general semantics to the world, *Science and Sanity*, and one of the pioneers of systems theory, Ludwig von Bertalanffy (1969), used the phrase *general system theory* (also known as general *systems* theory), following the example of general semantics (which followed the example of Einstein's general relativity theory). Korzybski himself died in 1950, just as systems theory was beginning to coalesce out of Norbert Wiener's

(1950, 1961) cybernetics, but his work influenced systems pioneers such as Buckminster Fuller (1971; Fuller, Agel, & Fiore,1970; Fuller & Applewhite, 1975), and Gregory Bateson (1972, 2002), and cybernetics and systems theory were central to the work of the Palo Alto Group that formed around Bateson, and that included the sociologist Erving Goffman (1959, 1961, 1963, 1967), anthropologists Edward T. Hall (1959, 1966) and Ray Birdwhistell (1970), and psychologist Paul Watzlawick and his colleagues (Watzlawick, 1976, 1983, 1988, 1990; Watzlawick, Bavelas, & Jackson, 1967; Watzlawick, Weakland, & Fisch, 1974); it also is very much at the heart of the work of physicist and philosopher Fritjoff Capra (1975, 1982, 1996, 2002).

McLuhan (2003) certainly interacted with many of these scholars, and brief mentions of systems concepts show up in *Understanding Media* via Kenneth Boulding's book, *The Image* (1956), but McLuhan's main body of scholarship, which was produced in the fifties and sixties, was completed by working in parallel to the development of systems theory in the sixties and beyond (but see McLuhan, 1966, for his take on cybernetics). So, McLuhan resonates with systems theory in his discussion of nonlinear processes, and the effects that occur when change is introduced into a system, but suffers from a lack of recourse to systems terminology and perspective. Walter Ong experimented with systems theory in his 1977 collection, *Interfaces of the Word*, and the evidence of a systems approach can also be discerned in the work on orality and literacy published by anthropologist Jack Goody (1977, 1986, 1987). But these efforts were not aimed at anything like an integration of systems theory into the study of media, technology, and symbolic form. Neil Postman (1970), however, incorporated systems theory in his early work on media ecology, drawing on Wiener, Bateson, Hall, Goffman, Birdwhistell, Watzlawick, Fuller, Boulding, Bertalanffy, and Ervin Lazlo (1972), as well as Korzybski and McLuhan. The systems view pervades the various addresses he gave on the topic of media ecology during the seventies, is very much a part of his first major single authored work, *Crazy Talk, Stupid Talk* (1976), and also comes into play in regard to the thermostatic view he introduces in *Teaching as a Conserving Activity* (1979). Christine Nystrom's doctoral dissertation completed under Postman's direction in 1973, and entitled *Towards a Science of Media Ecology: The Formulation of Integrated Conceptual Paradigms for the Study of Human Communication Systems,* brought the systems view together with media ecology in an attempt to form an integrated theoretical framework. Joshua Meyrowitz (1985) followed Nystrom's lead, and developed what he referred to as medium theory, integrating McLuhan and Goffman by defining both technological mediation and face-to-face communication as information systems. Meyrowitz does not highlight the fact that systems theory is the common ground by which he brings together two aspects of the field of media ecology, so that it serves as a hidden ground in his work, and to some extent for Kenneth Gergen (1991), who draws on Meyrowitz in applying Watzlawick's relational approach to the recent *fin de siècle* media environment. Systems concepts and approaches do appear in the media ecology literature over the past two decades (see, for example, Logan, 2007; Rushkoff, 1994, 2006; Strate, 2006; Zingrone, 2001), but the early efforts of Postman and Nystrom to integrate systems theory into the field of media ecology was to a significant extend abandoned.

Luhmann and Autopoiesis

Postman moved away from systems theory during the 1980s and 1990s, in part reflecting a more general trend in scholarship where the focus shifted away from the scientific and towards more humanistic, philosophical and critical approaches, and in part due to his own changing goals and interests. But at least in part, Postman and many others turned away from explicit engagement with the systems approach due to the limitations of systems theory itself, at least in what Katherine Hayles calls its first wave, in her book *How We Became Posthuman* (1999). It was in the second wave of systems theory, however, that the biologists Humberto Maturana and Francisco Varela (1980, 1992) introduced the concept of autopoiesis, which in turn was applied to social systems by Niklas Luhmann (1982, 1989, 1990, 1995, 2000a, 2000b). An autopoietic system creates itself by closing itself off from its environment, only letting in what it needs to maintain its existence. It establishes its boundary against its environment, and then limits what it allows to enter within the system. In this sense, the boundary or membrane allows the system to abstract part of what is in the environment into the system, abstracting nutrients and abstracting information. Luhmann draws on Korzybski in noting that social systems use language as a membrane, one that abstracts only the information that is needed, that is encoded in the appropriate symbolic form, and keeps out all the rest.

Luhmann also draws upon media ecology scholars Eric Havelock (1963, 1986), Walter Ong (1967, 1977, 1982), and Elizabeth Eisenstein (1979), in noting that the introduction of writing and printing technology increased the volume of information circulating within social systems, in turn allowing social systems to increase in complexity. Complexity is not simply a matter of a system growing larger; rather, it involves the development of specialized subsystems, each of which has its own boundary, maintained by a specialized language, and often involving a two-valued orientation. So for example, the legal system's boundary is based on guilt or innocence, the political system's boundary is based on winning or losing elections, and the educational system's boundary is based on having a degree or not having one. The mass media also function as a subsystem, and their main function is to provide the system with information about the outer environment, to create an inner map of the outer reality, but that picture of the world is obtained through the process of abstracting which inevitably results in vast simplifications, and distortions. This summary of Luhmann's perspective should help to explain why I have situated him as the common ground for and medium between Korzybski and McLuhan. I will now turn to a consideration of that common ground, and let the medium fade into the background as Korzybski and McLuhan's invisible environment.

Korzbyski and McLuhan

In the Preface to the Third Edition of *Science and Sanity*,[*] Alfred Korzybski (1993) wrote

> The origin of this work was a new functional definition of 'man' . . . based on an analysis of uniquely human *potentialities*; namely, that each generation may begin where the former left off. This characteristic I called the 'time-binding' capacity. Here the reactions of humans are not split verbally and elementalistically into separate 'body', 'mind', 'emotions', 'intellect', 'intuitions', etc.,

---

[*] The original work was published in 1933, and the Third Edition in 1947.

> but are treated from an organism-as-a-whole-in-an environment
> . . . point of view. (p. xxxxii)

In the Introduction to the Second Edition of *Understanding Media*,[*] Marshall McLuhan (2003) wrote

> The section on "the medium is the message" can, perhaps, be clarified by pointing out that any technology gradually creates a totally new human environment. Environments are not passive wrappings but active processes. (p. 12)

Korzybski and McLuhan both were highly independent, interdisciplinary, original, and unorthodox thinkers, and my aim is to draw on both of them, and on the approach that they hold in common. This approach, which they share with a number of other significant scholars and intellectuals, could be termed holistic, situational, contextual, relativistic or relational, cybernetic or a systems view, or ecological. Whatever the name that we give to it, the fundamental concern is with understanding what it means to be human; understanding ourselves as human beings, not in isolation, but in relation to our environments; understanding how we relate to our environments and how we *ought* to relate to our environments.

Ideally, we are concerned with how the organism-as-a-whole relates to the *total environment*, as McLuhan (2003) liked to put it. But in practice we may focus on a particular aspect of the environment, such as the cosmological or the geological, the physical or the chemical, the biological or the sociocultural. As an individual, the organism-as-a-whole exists in relationship to other individual human beings, in dyads and groups, in families and tribes, in organizations and associations, in cities and nations, in networks and in the global village. Every other organism-as-a-whole that the individual comes into contact with becomes part of the individual's environment. Of course, "organism" is a multiordinal term, referring to the individual, and also to the species-as-a-whole. The human species, which is one of many social species, is distinguished by our unique capacity for symbolic communication, which grants us our potential for time-binding. And the environment for our species-as-a-

---

[*] The original work published in 1964, and the Second Edition in 1965.

whole is nothing less than what Buckminster Fuller (1971) called *spaceship earth*.

Individually and collectively, the relationship between human beings and their environments is one that is fundamentally indirect. Externally, stimuli excite and irritate our sense organs and nervous systems. Internally, we construct a map of the environment out of the various excitations and irritations that we experience, a map that may be more or less structurally homologous with the outside world, but a map that is, simply stated, not the territory itself, as Korzybski famously put it. We therefore live in an inner environment, a perceptual environment, and a conceptual environment. And our relationship to the outer environment, being indirect, is therefore mediated, hence McLuhan's (2003) observation that the medium is the message.

This is not a solipsistic point of view, I hasten to add. Our concern is with the *relationship*, or if you like, the *interface* between the inner environment of the map, and the outer environment of the territory. It is along this interface that Korzybski (1993) made reference to verbal environments and semantic environments, and neurolinguistic and neurosemantic environments. It is along this interface that we can then speak of information environments, communication environments, symbolic environments, and technological environments. It is along this interface that McLuhan talked about media environments, and Neil Postman (1970) in turn defined media ecology as the study of media as environments.

Abstracting and Mediating

In discussing the ways that we relate to our environments, Korzybski employed the key term of "abstracting." He used the verb form of "abstracting" instead of the noun form of "abstraction" because he did not want us to think about abstractions as things, but rather as processes, and activities. As a verb, "abstract" can be defined variously as summarize, remove, separate, steal, and purloin. And what we as organisms do when we abstract is take into ourselves something from the outside environment. Unlike the processes of ingesting and inhaling, abstracting does not involve absorbing any material substance from the environment; what we abstract is information, which provides us with a necessarily incomplete and selective summary, or map of our environment.

On the Binding Biases of Time

All forms of life engage in some form of abstracting, in that all forms of life respond to stimuli. Organisms with nervous systems engage in the form of abstracting that we call sense perception. And the human organism, the time-binding species, also engages in the form of abstracting that we call symbolic communication, employing language and other symbol systems to generate and accumulate knowledge. Korzybski was not alone in this linking of perception and language. Susanne K. Langer (1957) argued that perception is a symbolic activity, a form of metonymy where the fraction of the environment that we take in stands for the environment as a whole. And McLuhan (2003) argued that language is a form of perception, indeed, that languages are organs of perception. And for Luhmann (1982, 1989, 1990, 1995, 2000a, 2000b), perception and language both contribute to the maintenance and functioning of the boundaries of self-organizing social systems.

The process of abstracting, as it encompasses the processes of perception and symbolic communication, allows us to do more with less, and therefore represents enormous efficiencies, especially for organisms with complex nervous systems such as ourselves. And efficiency, as Jacques Ellul (1964) has shown, is the basis of the technological imperative. Therefore abstracting, I would argue, is fundamentally associated with technological activity. And technological activity, I would suggest, can be traced back to the fact that all forms of life alter their environment, altering their environment to their own benefit, in theory if not always in practice, and altering their environment simply by their presence in the environment, and by their metabolism.

McLuhan used the term "technology" interchangeably with the term "medium" because all of our inventions and innovations are means by which we relate to our environment, by which we mediate and interface with our environment. Our technologies and techniques help us to abstract information from our environment; help us to accumulate, share, and preserve knowledge; help us to communicate and commune with one another; and help us to act upon our environment and alter it for better or for worse. We can study an individual medium or technology, its unique characteristics, biases, and effects. And we can also study the media-environment-as-a-whole, recognizing that the individual medium does not exist in isolation, but in complex interaction with other media. But going beyond the media-as-environments point of view, I

60

want to suggest that McLuhan's approach can best be understood when we employ the verb form of mediating. Along these lines, the French media ecologist Regis Debray (1996), drawing on the field of *semiology*, which is concerned with the process of *signification*, calls his field of study *mediology*, which is concerned with the process of *mediation*. Admittedly, there is some potential for confusion with the ways in which the terms mediation and mediating are used in the legal sector, but there is also some benefit from associating media with activities such as negotiation.

With this way of understanding media, we can say that we relate to our environments through a process of mediating, of interfacing, of abstracting. From the systems perspective (Bateson, 1972; Bertalanffy, 1969; Laszlo, 1972, 1996), we can only take in part of our environment through abstracting because its totality would overwhelm us. All systems must maintain boundaries with their environments in order to establish and maintain their integrity as systems. Indeed, as previously noted, Maturana and Varela (1992) argue that it is only by closing itself to its environment to a significant degree that a system can organize itself, that is, that an independent system can come together as a system in the first place (again, this has been insightfully elaborated upon and applied to social systems by Luhmann, 1982, 1989, 1990, 1995, 2000a, 2000b). We create barriers for our own protection, biologically, psychologically, and sociologically. And we do so technologically as well, or as Max Frisch (1959) observed, "technology is the art of never having to experience the world" (p. 178).

McLuhan regarded media and technology as extensions of the human organism, following a tradition that can be traced back to Edward T. Hall (1959), C.K. Ogden and I.A. Richards (1923), and Ralph Waldo Emerson (1883). But McLuhan insisted that every extension is also an amputation. The medium that extends our reach into the world does so by situating itself between ourselves and the world, so that it also becomes a barrier between ourselves and the world. And as a barrier, the medium becomes part of our world, part of our environment. In sum, as we relate to our environment, we reject as well as select. We filter. We mediate. Or as I like to say, the medium is the membrane. We dance along the edge of chaos and order, opening and closing, extension and amputation, the external and the internal.

The Mode of Abstracting

Both Korzybski and McLuhan were concerned with *differences that make a difference*, as Gregory Bateson (1972) would put it. But their point of view on these differences were somewhat... different. Korzybski, being an engineer, and of a scientific and mathematical bent, looked at the process of abstracting along a vertical axis of higher and lower levels. Sense perception constitutes the lowest level of abstraction, symbols a higher level. Images are less abstract than words. Names that are attached to an individual are less abstract than labels that refer to an entire category. Following Korzybski's lead, we can say that television is less abstract than a book. And we can say that the written word is more abstract than the spoken word because, as Walter Ong (1982) explains, writing is a secondary symbol system that represents speech, our primary symbol system, in a visual form.

McLuhan, being a scholar of English literature, focused on qualitative rather than quantitative differences. With this in mind, we can add to the vertical axis of level of abstracting a horizontal axis that we can call *the mode of abstracting*. For example, television and movies are different media, and therefore represent different methods of abstracting. One is not particularly more or less abstract than the other, they are just qualitatively different modes. The same could be said of sound recordings and radio, or a magazine and a newspaper, or a parchment manuscript and papyrus scroll, or hieroglyphics and cuneiform, or a dialogue and a public address. Following the insights of Edward Sapir (1921), Benjamin Lee Whorf (1956), and Dorothy Lee (1959), we can also understand that different languages (e.g., English, Spanish, French, Hindi, Mandarin, Hopi), represent different modes of abstracting, and mediating. And the same can be said of different forms of sense perception. As McLuhan made clear, vision represents a different mode of abstracting than hearing, or touching, or smelling or tasting. And senses trained in different ways represent different modes of abstracting, so that literates, for example, use their eyes differently than nonliterates do, and this in turn alters the way that all of the other senses are used in concert.

Taking into account the mode of abstracting as well as the level of abstracting gives us a better handle on the process of abstracting, and the process of mediating. We mediate with our bodies, our sense organs and nervous systems. We mediate through our languages, art forms, and symbol systems. And we mediate through our technologies, techniques, and technical systems. We relate to our environment, not as separate entities, but as interdependent parts of an ecosystem. What that means is that the organism-as-a-whole is influenced and shaped by its environment, which is, as McLuhan noted, not a passive wrapping but an active process. And the environment-as-a-whole is influenced and shaped by the organisms that are a part of it. We have changed our environments radically through our inventions, through our ideas, and through our activities, sometimes for the better, sometimes for the worse. And in our efforts to make things better, we certainly would do well to remember Margaret Mead's inspiring words: "Never doubt that a small group of thoughtful, committed citizens can change the world. Indeed, it is the only thing that ever has."[*] But to change the world for the better, we must understand the world, understand how we relate to the world, how we can change the world and what the consequences of change might be.

What we need then is an ecology of mediating, an ecology of abstracting, an ecology of knowing. We need an ecology of knowledge, by which I mean both knowledge in the academic sense, and know-how in the everyday sense, knowledge that is both theoretical and practical, pure and applied, both theory and praxis. We need an ecology that incorporates both form and technology, the inner landscape and the outer environment, the map and the territory. We need an ecology that is built upon the relationships between systems and their environments, as they are mediated by their boundaries. In other words, we need an ecology of Korzybski, Luhmann, and McLuhan.

---

[*] There is no published reference for this quote—see http://www.interculturalstudies.org/faq.html#quote for a discussion of its oral origins. For more on this theme, see Mead (1964).

Chapter Four
On the Binding Biases of Time

have chosen to address the theme of time because it is a topic that I find fascinating, and I would discuss it in great depth and detail if I could, but in the words of the Roman poet Virgil, *tempus fugit*. Or as Groucho Marx put it, *time flies like an arrow, fruit flies like a banana*. That quip demonstrates that syntax and meaning is subject to the principle of relativity, just as much as time and gravity. What this means is that it is not just time that is relative, but that our ways of representing time, talking about time, telling time as it were, are equally relative. So, for example, Stephen Hawking writes in his bestselling book, *A Brief History of Time* (1998), that the universe started off with a bang about thirteen or fourteen billion years ago, and is continuing to expand today. Whether the expansion will go on forever, or reach a limit, and reverse itself, contracting back down to a single point, remains a matter of some debate. Either way, the fact remains that the Big Bang was an explosion so massive that it is still going on, some thirteen or fourteen billion years later, with no end in sight. The explosion is taking place on such a vast scale that we do not experience it as an explosion, but we are all riding the Big Bang, clinging to a tiny bit of debris that we call Earth, as our galaxy moves at a rate of 300 kilometers or 185 miles per second.

The Bible tells us, "To every thing there is a season, and a time to every purpose under heaven," to which Pete Seger added the refrain, "turn, turn, turn." We traditionally looked to the cycles of nature for our sense of time, and tried to capture those cycles, in turn, in our calendars and clocks. But we also have a sense of time as an irresistible forward motion. It was not until the nineteenth century that physicists established the Second Law of Thermodynamics, which states that the universe has a statistical tendency to move towards a greater amount of entropy over time, meaning that the passage of time is irreversible; this is sometimes referred to as *time's arrow*. But the notion of history as an unfolding progression dates back to antiquity, and no doubt our prehistoric ancestors understood the process of aging, and the passages from birth to childhood to maturity to death. Having turned fifty-two not long before I began writing this essay, I note with some resentment that in English I have to say that *I am* fifty-two years old, fifty-two years *old*! I much prefer the French language in this instance, as then I would say, *j'ai cinquante-deux ans*, I *have* fifty-two years. Now, Noam Chomsky would say that they mean the same thing in their deep structure, but fortunately we are in the post-Chomsky era now, and can safely say that the two statements reflect quite different views of time. In English, you become old, it is a change of state; being fifty-two, I am no longer fifty-one or fifty, I have lost those ages in the process of metamorphosis. In French, you gain years, it is an accumulation; having fifty-two years, I did not lose age fifty-one or fifty, or any other age—I still have them! They represent the experience, the knowledge, and presumably the wisdom that I have gained.

Marshall McLuhan talked about how we move into the future looking into the rearview mirror (McLuhan & Fiore, 1967). That is an automobile metaphor, and I found it instructive to learn that he never actually drove very much. But even if his metaphor is a bit askew, his point is quite valid, that we tend to live in the past, because that is all that we know. McLuhan pointed out that a prophet is not someone who foresees the future, but rather someone who tells you what is happening right now. In military circles, it is commonly said that there is a tendency to prepare to fight the last war you were in, instead of preparing for the war you are actually about to fight. And it seems to me that part of the problem is that we think of time in terms of space, imagining time as a line or a road that we are traveling on, moving forward into the future. McLuhan reminds us that we can see nothing of the future that lies before us, while the past is laid out clearly for our inspection. In this sense, then, we are walking backwards into the future, a metaphor employed in some tribal cultures, notably the Maori.

I think it important to keep in mind that the road is only a metaphor for thinking about time, it is not the phenomenon itself. We think of time as one-dimensional, a line, but why not two dimensions of time, or three dimensions, like the length, breadth, and width of space? We could alternately imagine time as a balloon being filled with air, continually expanding in size. Thinking of time as visual space, whether linear or multidimensional, suggests the possibility of time *travel*, but being only a metaphor, I suspect that notions of going back to the future and forward into the past will remain confined to the realm of science fiction and fantasy. Instead of thinking of time visually, why not imagine time as a purely sonic phenomenon, as a conversation or a piece of music. Indeed, our sense of hearing is intimately linked to our sense of time, for as Walter Ong (1982) puts it, "sound exists only when it is going out of existence" (p. 32). We can even think of time through the sense of smell, and in fact McLuhan notes in *Understanding Media* (2003) that research on the brain indicates that smell is closely associated with memory; he also notes that up until missionaries introduced the mechanical clock to China and Japan in the seventeenth century, incense had been used for millennia in East Asia to measure time's passage, and distinguish special occasions.

A Double Blind of Binding Biases

At this point, having neither world enough nor time, I must put an end to this meandering introduction, and begin in earnest by quoting from James W. Carey's highly respected book, published in 1989, *Communication as Culture*, from an essay about Harold Innis:

> Innis argued that changes in communication technology affected culture by altering the structure of interests (the things thought about), by changing the character of symbols (the things thought with), and by changing the nature of community (the arena in which thought developed). By a space-binding culture he meant literally that: a culture whose predominant interest was in space—land as real estate, voyage, discovery, movement, expansion, empire, control. In the realm of symbols he meant the growth of symbols and conceptions that supported these interests: the physics of space, the arts of navigation and civil engineering, the price system, the mathematics of tax collectors and bureaucracies, the entire realm of physical science, and the system of affectless, rational symbols that facilitated those interests. In the realm of communities he meant communities of space: communities that were not in place but in space, mobile, connected over vast distances by appropriate symbols, forms and interests.
>
> To space-binding cultures he opposed time-binding cultures: cultures with interests in time—history, continuity, permanence, contraction; whose symbols were fiduciary—oral, mythopoetic, religious, ritualistic; and whose communities were rooted in place—intimate ties and a shared historical culture. The genius of social policy, he thought, was to serve the demands of both time and space; to use one to prevent the excesses of the other:

to use historicism to check the dreams of reason and to use reason to control the passions of memory. But these were reciprocally related tendencies. As cultures became more time-binding they became less space-binding and vice versa. The problem again was found in dominant media of communication. Space-binding media were light and portable and permitted extension in space; time-binding media were heavy and durable or, like the oral tradition, persistent and difficult to destroy. In propositional form, then, structures of consciousness paralleled structures of communication. (pp. 160-161)

Those of you who are familiar with Carey's scholarship know that he was a leading expert on the work of the Harold Innis, and that Innis was a Canadian economist who, towards the end of his career, turned his attention to the study of communication, and particularly the relationship between forms of communication and social organization. Carey stood alongside Marshall McLuhan in acknowledging the great debt that we owe Innis for his pioneering investigations into the study of media, and Carey's insights into Innis's scholarship were second to none. But, those of you who are familiar with Harold Innis's classic work, *The Bias of Communication* (1951), may have noticed something curious about Carey's explication. As the title of his book indicates, Innis argued that different modes of communication are characterized by different inherent biases, an idea that has come to be foundational for the field of media ecology. And Innis specifically contrasted technological and cultural biases towards space and towards time. But Carey, rather than using Innis's terms, *space bias* and *time bias*, speaks of space-*binding* and time-*binding*. It was a seemingly minor and harmless substitution, to be sure, except for the fact that the phrases space-binding and time-binding are established terms in the discipline of general semantics, having been coined by Alfred Korzybski in his first book, *Manhood of Humanity* (1950), and included in his magnum opus, *Science and Sanity* (1993).

Not long after I was appointed Executive Director of the Institute of General Semantics, someone on a general semantics online discussion group came across Carey's quote and asked why these terms were being used without any attribution to Korzybski. I explained that there probably was some mistake, that Innis used the term *bias*, not *binding*, that *time bias* referenced concepts significantly different from *time-binding*, and as far as I could tell, Innis did not draw upon Korzybski's work at all. On further reflection, though, I think it reasonable to assume that Innis was aware of Korzybski's work, as most North American intellectuals were in the mid-twentieth century, including Lewis Mumford, one of Innis's influences. Along similar lines, Carey does not make any reference to Korzybski in his writings, as far as I know, but I am certain that a scholar of intellectual history of Carey's caliber was familiar with Korzybski's work and terminology. So in the end, I cannot say whether Carey meant to draw a connection between Innis and Korzybski, or substituted the terms solely for stylistic reasons, or simply made a mistake. But that point of either conflation or confusion gave me the idea to draw on both terms, and entitle this essay, "On the Binding Biases of Time." And my intent is to address the subject of time within it, at least as much as time, and space, permit.

As human beings, we are both blessed and cursed with a consciousness of time that is unique among the myriad forms of life, at least as far as we can tell. Granted, all forms of life function within time, seek to maintain their existence over time, and do so by responding to certain changes in their environment, changing in some way to meet the demands of a dynamic environment that itself changes over time, and also acting upon their environment in an attempt to alter it in ways conducive to their own survival. One of the defining characteristics of life forms is that they attempt to produce copies of themselves over time, to ensure the survival of their species, or as Richard Dawkins (1989) argues, the gene itself is a replicator, replication is its only imperative, and this drives all other biological and behavioral processes. Every characteristic that is typically used to define the somewhat nebulous concept of life involves activity occurring over time, for example metabolism, homeostasis, response to stimuli, growth, adaptation, and reproduction. We might therefore distinguish life from non-life by the organism's sense of time, which might be called an awareness or consciousness of time depending on how you choose to define those terms. What makes an organism alive is its active engagement with time, its use of time to relate, adjust, and modify its environment, and its use of time to relate, adjust and modify its own self. This view gives new meaning to the term *time-life*.

As J.T. Fraser (1987), the founder of the International Society for the Study of Time, puts it, "to say that a living organism is an orchestra with trillions of instruments that are kept playing in a coordinated fashion from instant to instant is, of course, a metaphor. But it is a useful and appropriateone, because it reminds us of the organic solidarity that constitutes the life process" (p. 128). Fraser goes on to argue

> The literature of biological clocks regularly asserts that the reason they evolved was to help the organism survive. This is as inadequate a view as if I claimed that the musicians of an orchestra help the orchestra make music. Musicians do not help the orchestra, they *are* the orchestra. Likewise, biological clocks do not help a living organism survive, they *are* the organism. (pp. 131-132)

Further, the sense of time that distinguishes the organic from the inorganic extends beyond the present. Once more, as Fraser explains

> Some 3.5 billion years ago, as perceived and measured by us clock watchers, self-organizing systems characterized by a know-how for defining an organic present came into being. Biogenesis created a new kind of time, one that was more advanced, more evolved, than the time of the physical world: expectation and memory created new categories of time in terms of the organisms' self-interests. Nothing in the physical world ever remembers anything, not even a memory circuit does. Nothing in the physical world ever expects anything, not even an alarm system. Only for living organisms do future, past, and present constitute reality. (p. 136)

While life in general exists within an environment of memory, expectation, and perception, species differ in the ways in which they organize and access their memories, and in their ability to anticipate alternatives and plan for contingencies. And of all the myriad species that populate the planet, none come close to the unique time consciousness of the human race. It is not just that, as compared to other forms of life, we human beings have more powerful memories and stronger cognitive abilities that enhance our ability to anticipate and plan. Rather, our sense of time represents a quantum leap beyond any kind of time consciousness that has existed before us. The qualitative difference extends so far as to make us the only species that can develop an awareness of our mortality. Ernest Becker argues that this awareness is so devastating to our self-esteem that we require extraordinary means for coping with that awful knowledge. According to Becker (1971, 1973), the answer to that problem comes through culture, which provides us with the means for living as heroes within our own narrative, and by seeing ourselves as living heroically, provides us with the means for the denial of death.

The prospect of our own demise renders human time consciousness at times poignant, at times courageous, at times absurd, arrogant, and grotesque. We may be consoled, however, by the realization that others will carry on in our absence, that we will be remembered by others just as we remember those who are no longer with us, that activities that we have begun can be taken up and continued by others, that plans that we construct may be brought to fruition by others. We call certain documents a *will*, a *last will and testament*, and a *living will*, because they are in fact extensions of our consciousness reaching into a future where we are no longer present or mentally active; they are attempts to project our intentions and determinations, our decisions and desires, our *will* beyond the lifespan of our consciousness. Beyond memory and will, we may find some consolation in the realization that something of what we have learned in life, some aspect of our awareness, some element of our thought processes, some portion of our ways of knowing, some subset of the knowledge that we have accumulated has been passed on to others, and will survive our own individual demise.

Korzybski's Concept of Time-Binding

I want to suggest to you, however, that this sort of acute time consciousness, while intensified by our awareness and understanding of death, emerges as a consequence of what Korzybski referred to as time-binding. The basic idea of time-binding is by no means a radical notion. It is the idea that human beings make progress from one generation to the next by virtue of our ability to preserve and transmit knowledge. Time-binding is an elaboration on the famous quote by Isaac Newton: "If I have seen a little further it is by standing on the shoulders of giants." The quote itself is a product of time-binding, as earlier variants can be traced back to at least the twelfth century. Nowadays, we have grown uncomfortable with the word *progress*, so we are more likely to talk about evolution, for example, in reference to cultural evolution. If you really think about it, though, in this instance evolution is being used to a large extent as an euphemism for progress. At one time, the talk was of evolution to a higher state of being; more recently we speak of evolution towards greater complexity. And while I understand the need to avoid the triumphalism associated with the concept of progress in the early twentieth century, our language has grown poorer and less precise for having eliminated that word from our working vocabularies.

While Korzybski introduced the term time-binding to signify a radical break between human beings and other forms of life, it is also possible to view time-binding as the product of an evolutionary process. In other words, we might say that we are not the only species to engage in some form of time-binding activity, instead observing that other animals are capable of imitation and learning, thereby passing on useful behaviors from one generation to the next. Thus, time-binding has slowly evolved from forms of life that are entirely dependent on the self-replication of DNA to maintain their existence over time, to species that are capable of engaging in increasingly more elaborate forms of social learning. For most animals, this limited form of time-binding allows for adaptation of behaviors in response to a dynamic, changing environment. If you take the view of some anthropologists that animals as well as humans can be said to have culture, this sort of cultural evolution is directed towards homeostasis, rather than progress. Indeed, we might say that much of human cultural evolution is directed towards homeostasis, especially in tribal societies. What this means is that human beings have the potential to make progress through time-binding, but we do not necessarily realize that potential. In fact, for most of our history, we have not.

Korzybski used the concept of time-binding as the basis of his definition of the human race as a unique class of life, in contrast to animals, which he referred to as space-binding, and plants, which he termed chemistry-binding. This three-fold schema, while presented as science, seems inadequate even by the standards of early twentieth century biology. What then to make of it? One possibility is to view it as a metaphor, a poetic or rhetorical device. Or it can be seen as a heuristic device, a way to start thinking about humanity and what it might take to solve problems like war, violence, and oppression. But Korzybski's classifications can be better understood when we take into account the fact that his background was in engineering. Engineers are concerned with pragmatic questions and practical concerns, which is no doubt why this Polish nobleman found a following, and a home, in the United States. As a form of applied science, engineering is concerned above all with getting specific tasks accomplished, with work. From the point of view of physics, work requires the application of force, and force is the product of energy.

Engineering, then, is all about energy, and it is worth noting that our contemporary understanding of energy was relatively recent when Korzybski began his investigations. As Thomas Kuhn explains in his paradigm-shifting work, *The Structure of Scientific Revolutions* (1996), the pioneers of electrical research viewed electricity as a substance. In particular, these *electricians*, as scientists such as Benjamin Franklin referred to themselves, believed that electricity was a fluid, and to this day we use terms such as *flow* and *current*. Along the same lines, light was once thought to be a form of matter, be it particle, medium, or emanation, and from antiquity fire was thought to be one of the elements. It was not until the nineteenth century that the idea of *vis viva*, of living force, was finally replaced by that of energy, and the laws of thermodynamics were formalized. And at the beginning of the twentieth century, Albert Einstein introduced his famous equation, $E = MC^2$, which establishes that energy and matter are essentially equivalent, the third element in that equation being the square of the speed of light, which is a measure of time. What all this represents is a paradigm shift away from the view that the universe can be best understood as matter, to a view that the universe is essentially energy, and matter is simply a form of very slow and stable energy, or potential energy. It is a shift away from viewing "things" as static and substantial, and towards viewing all phenomena as dynamic processes. In the mid-nineteenth century, Karl Marx and Fredrich Engels (2007) argued that as a result of the growth of capitalism and industrialism, "all that is solid melts into air" (p. 14), but this is an apt description of the scientific revolution that was going on at the time that they wrote their *Communist Manifesto*, a paradigm shift that McLuhan (1962, 2003) noted was preceded and precipitated by the development of electrical technology, notably the telegraph.

As an engineer working in the enthusiastic wake of a scientific revolution, Korzybski's theory of time-binding was all about energy; indeed, he was an ardent admirer of Einstein, and in fact called his early work a *general theory of time-binding*, following the example of Einstein's *general theory of relativity*. And Korzybski's theory begins with the sun as a source of energy for life on earth. And more than any other form of life, plants have evolved a way to capture and store that energy, which is why he called plants the chemistry-binding class of life. That stored energy is then used by animals, who convert it into motion, that is, kinetic energy, moving freely about in their environment in ways that plants are not capable of, and that is why he called animals the space-binding class of life. Human beings are able to use that energy to move through space as well, but we have also found a way to store it, not chemically, but in the form of knowledge, which makes us the time-binding class of life. Korzybski also regards these categories in terms of dimensionality, chemistry-binding being one dimensional, space-binding adding a second dimension, and time-binding give us humans sole access to a third dimension of life. Here too we can see the influence of Einstein, who posited time as a fourth spatial dimension, better understood as the unified phenomenon of spacetime. But dimension, after all, is just a fancy way of saying measure, and Korzybski's three dimensions of binding are measures of energy. And energy, once again, is what is required to get work done. This leads to a rather interesting economic commentary that can be found in *Manhood of Humanity* (1950):

> The potential use-values in wealth are created by human work operating in time upon raw material given by nature. The use-values are produced by time-taking transformations of the raw materials; these transformations are wrought by human brain labor and human muscular labor directed by the human brain acting in time. The kinetic use-values of wealth are also created by human toil—mainly by the intellectual labor of observation, experimentation, imagination, deduction and invention, all consuming the precious time of short human lives. It is obvious that in the creation of use-values whether potential or kinetic, the element of time enters as an absolutely essential factor. The fundamental importance of time as a factor in the production of wealth—the fact that wealth and the use-values of wealth are

literally the natural offspring of the spiritual union of time with toil—has been completely overlooked, not only by the economics, but by the ethics, the jurisprudence and the other branches of speculative reasoning, throughout the long period of humanity's childhood. In the course of the ages there has indeed been much "talk" about time, but there has been no recognition of the basic significance of time as essential in the conception and in the very constitution of human values. It is often said that "Time is Money"; the statement is often false; but the proposition that Money is Time is always true. It is always true in the profound sense that Money is the measure and symbol of Wealth—the product of Time and Toil—the crystallization of the time-binding human capacity. IT IS THUS TRUE THAT MONEY IS A VERY PRECIOUS THING, THE MEASURE AND SYMBOL OF WORK— IN PART THE WORK OF THE LIVING BUT, IN THE MAIN, THE LIVING WORK OF THE DEAD. (pp. 116-117)

In Korzybski's analysis, wealth is for the most part not something that people earn on their own, a view that runs contrary to the American dream, our Horatio Alger tales of going from rags to riches, and our glorification of the entrepreneur. Rather, Korzybski reminds us that most wealth is inherited. Unlike Marx, Korzybski is not concerned with the fact that inherited wealth tends to stay in the hands of a small population over time, generally within families and socioeconomic classes. Instead, Korzybski focuses on wealth as a common human inheritance, which should in turn be utilized for the common good, rather than private gain; it should be noted as well that his concept of wealth consists of more than just money and material goods, but also and especially knowledge and knowhow. In remarking on "the capitalist era" he states

It may seem strange but it is true that the time-binding exponential powers, called humans, do not die—their bodies die but their achievements live forever—a permanent source of power. All of our precious possessions—science, acquired by experience, accumulated wealth in all fields of life—are kinetic and potential use-values created and left by by-gone generations; they are humanity's treasures produced mainly in the past, and conserved for our use, by that peculiar function or power of man for the binding of time" (p. 119).

Essentially, then, every invention, every innovation, every human advancement is the product of generations, indeed millennia of previous discoveries, tens of thousands of years of intellectual and physical labor. Thus, Korzybski comments

> This fact, of supreme ethical importance, applies to all of us; none of us may speak or act as if the materia sul or spiritual wealth we have were produced by us; for, if we be not stupid, we must see that what we call our wealth, our civilization, everything we use or enjoy, is in the main the product of the labor of men now dead, some of them slaves, some of them "owners" of slaves. The metal spoon or the knife which we use daily is a product of the work of many generations, including those who discovered the metal and the use of it, and the utility of the spoon.
>
> And here arises a most important question: Since the wealth of the world is in the main the free gift of the past—the fruit of the labor of the dead—to whom does it of right belong? The question can not be evaded. Is the existing monopoly of the great inherited treasures produced by dead men's toil a normal and natural evolution?
>
> Or is it an artificial status imposed by the few upon the many? Such is the crux of the modern controversy. (p. 124)

Korzybski's critique of capitalism and commercialism was not an affirmation of communism or socialism, however, but rather the basis for arguing for government based on scientific principles, a technocracy run by individuals involved in *human engineering* (as he put it), and a society where everyone would employ a rational, scientific approach in every aspect of their lives. From a contemporary perspective, this sounds at best naïve and idealistic, if not ominous and threatening, but I think it important to recall how differently we viewed science, technology, engineering, and progress in the early twentieth century. Korzybski's optimism was paralleled by that expressed by Thorstein Veblen in the *The Engineers and the Price System*, also published in 1921, and in Lewis Mumford's hopeful view of the transformative potential of electrification in his 1934 tome, *Technics and Human Civilization*. Politics aside, what is of great significance is that Korzybski differentiates between different types of time-binding. Just as earlier I suggested that animals are capable of a form of time-binding that is largely homeostatic rather than progressive, Korzybski argues that human time-binding progresses slowly for the most part, arithmetically at best, except for the advancements that are made in science, technology, and engineering, where time-binding becomes rapid, and progress geometric.

The disconnect between progress in science and technology on the one hand, and ethics and human relations on the other, is a subject that many have commented on. Korzybski's proposal was to apply the scientific method, which has proven to be so extraordinarily successful in enabling us to predict and manipulate various aspects of our environments, to all aspects of human life. Upon further investigation, he came to understand that what set human time-binding apart from animal behavior so very dramatically was the human capacity for language and symbolic communication (Korzybksi, 1993). Language is a storage medium, and the language that we speak is not our own invention, but the product of untold generations that have gone before us. Following the Sapir-Whorf Hypothesis, the language that we are taught is a medium that transmits the stored perceptions and worldview of the past (see Sapir, 1921; Whorf, 1956). Expressed in terms of energy, language is a battery, it stores cognitive energy that we use to get work done, and we recharge that battery by teaching our young how to speak. It follows then that differences in the way that we use language can lead to differences in the process of time-binding. Thus, Korzybski concluded that the ways in which scientists and engineers use language in their professional activities are much more effective than the imprecise and ambiguous way that language is used otherwise. Consequently, he developed general semantics as a means of extending the scientific approach to all of communication, perception, and evaluation, and thereby improving the efficiency of time-binding and increasing the rate of progress in all areas of human activity.

The process of time-binding, then, is subject to change, evolution, and progress; in short, time-binding is subject to time-binding. There is an implied theory of history in *Manhood of Humanity*, a history of human development that is, once again, broken up into three stages. First, there is the pre-scientific stage, where time-binding exists, but progress is very slow. Korzybski has relatively little to say about this stage, and perhaps he might have referred to this as the infancy of humanity (which is not to say that I support such a characterization). Second is the scientific stage, where the adoption of scientific method allows for rapid progress in specialized sectors such as applied science, technology and engineering, and in our knowledge of pure science and mathematics, but not in any other aspect of human affairs. This is what Korzybski means by the childhood of humanity. The third stage would be when all human life is informed by and governed by a scientific approach. I think we could refer to this period as post-scientific, not because science would be obsolete, but simply because it would become ubiquitous and environmental; this follows the same logic in which Fredric Jameson (1991) explains that the postmodern is considered the period following modernization, in which the process of modernization has been completed and is no longer an issue. This period would be the "manhood of humanity," our mature phase, a stage he believed we were about to enter.

Innis's Notion of Time Bias

Korzybski was wounded as a Polish soldier in the Russian army during the First World War, and went on to publish *Manhood of Humanity* in 1921, *Science and Sanity* in 1933, and found the Institute of General Semantics in Chicago in 1938. Harold Innis was wounded as a Canadian soldier in the British army during the First World War, earned his PhD from the University of Chicago in 1920, and went on to teach at the University of Toronto, where he became Canada's leading economist. The parallels are interesting, but as far as I can tell, they never met or communicated with one another. Innis published several books on the subject of Canada's political economy during the twenties, thirties, and forties, and did not turn his attention to the study of communication until after the Second World War. It was not until 1950, the year that Korzybski died, that Innis published *Empire and Communications* (1972), followed the next year by *The Bias of Communication* (1951), and then by *Changing Concepts of Time*, published in 1952, the year that Innis died. And it was in *The Bias of Communication* in particular that Innis discussed the biases of time, and space. Whereas Korzybski was concerned with the question of what distinguishes humanity from other forms of life, Innis was concerned with the question of what distinguishes one type of human society from another. And whereas Korzybski brought an engineer's concern with work and energy to the study of time, Innis brought an economist's concern with raw materials and staples; if time is energy to Korzybski, the media by which we communicate over time is akin to coal and oil to Innis.

Korzybski studied time, and that led him to the study of communication. Innis studied communication, and that led him to the study of time. Communication, however, has typically been talked about in terms of transportation, transmission, or pipeline metaphors. Generally, communication is seen as being a process of moving messages from point A to point B, or from point A to many different point Bs in the case of mass communication. As such, the focus is on how far a message can travel, how fast it could traverse a given distance, and how widely it can be disseminated. In other words, the preoccupation has been with communicating over space. It therefore represents a significant breakthrough on Innis's part to realize that communication can take place over time as well as over space. Building on Innis, Carey came to stress the role of communication in the formation and preservation of communities, including the imagined communities that we call nations, and in the maintenance of social cohesion and cultural continuity. Carey (1989) called it the ritual view of communication, which he contrasted with the transportation view, and which implies that we ought to focus pay attention to the process of communing as well as commuting.

Innis, then, came to his own understanding of what Korzybski termed time-binding, but he moved in a different direction, observing that in the process of binding time, we bind ourselves together in social units, as families and tribes, communities and cities, nations and societies. And as we bind ourselves together in this way, we ourselves become bound by time, time becomes the ties that bind, and we become prisoners of our remembered past, and imagined future. And as we bind ourselves together in different ways, as the means by which we bind time changes, so too does the character of human culture. This is central to Innis's insight, and is part of a broader generalization that differences in the way that we communicate with others and with ourselves, differences in the way that we mediate between ourselves and our environment, are what Gregory Bateson (1972) would refer to as *differences that make a difference*; they are systemic differences that have a powerful influence on the way that we think, feel, and perceive the world; on our consciousness, identity, and relationships; on our forms of social organization and our culture.

In *The Bias of Communication* (1951), Innis states

> My bias is with the oral tradition, particularly as reflected in Greek civilization, and with the necessity of recapturing something of its spirit. For that purpose we should try to understand something of the importance of life or of the living tradition, which is peculiar to the oral as against the mechanized tradition, and of the contributions of Greek civilization. (p. 190)

Innis goes on to argue

> The oral dialectic is overwhelmingly significant where the subject-matter is human action and feeling, and it is important in the discovery of new truth but of very little value in disseminating it. The oral discussion inherently involves personal contact and a consideration for the feelings of others, and it is in sharp contrast with the cruelty of mechanized communication and the tendencies which we have come to note in the modern world. The quantitative pressure of modern knowledge has been responsible for the decay of oral dialectic and conversation. (p. 191)

Innis favored oral tradition for its flexibility, but also understood its limitations, as he also notes, "an oral tradition implies freshness and elasticity but students of anthropology have pointed to the binding character of custom in primitive cultures" (p. 4). This of course is another example of the binding character of time-binding, but it is also important to acknowledge that at the time that Innis was writing, he himself was bound or shall we say constrained by the fact that orality-literacy studies were relatively new and undeveloped. Still, they were far from unknown, as Milman Parry's (1971) groundbreaking research establishing the basic characteristics of oral composition and oral culture took place during the 1920s and 1930s, at the same time that Korzybski was writing *Manhood of Humanity* and *Science and Sanity*. And Innis had the benefit of a brief but fruitful exposure to the renowned classicist Eric Havelock (1963), before Havelock left Toronto for Yale University. Indeed, it might be said that two fundamental and parallel discoveries occurred during this period, one concerning the understanding of time in contrast to space, and the other in regard to the understanding of sound in contrast to vision. The two go hand in hand, sound being a dynamic phenomenon, only existing as it goes out of existence, so that it cannot give the illusion of stopping or otherwise stepping outside of time in the way that vision does. Pause a video and you get a freeze frame, but pause an audio recording and all you get is silence. The relationship between time and sound perhaps sheds new light on the significance of Einstein's violin playing, and certainly represents a cornerstone of the media ecology perspective.

Sound being ephemeral, time-binding in oral cultures is entirely dependent on human memory, not individual memory alone but collective memory. Moreover, memory is not a thing, not a substance, but a form of energy, the activity of remembering; but more than that, memory is a performance, an active process of commemoration (Hobart & Schiffman, 1998). To be kept in collective memory, knowledge becomes attached to dramatic narrative, conveyed in the form of extraordinary agents performing remarkable actions, and typically expressed in mnemonic forms such as poetry, songs, and sayings. The singer of tales in an oral culture, having no written text to study, does not have the concept of verbatim memorization that we literates do, so that no two oral performances are alike; in fact, the singer is quite willing to vary the performance to accommodate the situation, mood of the audience, and other factors. The multiformity of oral performance is the key to the flexibility of oral tradition, as the tradition being fluid can easily adapt to meet changing circumstances (Lord, 1960; see also Havelock, 1963; Ong, 1982).

To give an example, the twelve tribes of ancient Israel are represented in the Bible by the twelve sons of Jacob, each of whom carries the name and is presented as the ancestor of one of the tribes, and this is a common motif in oral cultures. When the Assyrians destroyed the northern kingdom of Israel, ten of the twelve tribes disappeared, their people presumably killed, enslaved, or assimilated. But because the story was part of a written tradition, the ten lost tribes were not forgotten, and Jews and Christians alike searched the world for them; upon the discovery of the New World, some thought that the Native Americans might be the descendents of the ten lost tribes. By way of contrast, Jack Goody and Ian Watt (1968) relate the story of a West African people who told the tale of seven brothers, each the ancestor of a neighboring tribe. British researchers recorded this myth early in the twentieth century, and no subsequent studies were carried out until sixty years later. During that time, two of the tribes had disappeared, and the myth had changed accordingly, so that they now told the story with only five brothers instead of seven. Not only was there no acknowledgement that any change had occurred on the part of these peoples, but they insisted that this was the story that they had always told. Goody and Watt refer to this characteristic of oral cultures as *homeostatic*, and Walter Ong (1982) comments on this, saying "oral societies live very much in a present which keeps itself in equilibrium or homeostasis by sloughing off memories which no longer have present relevance" (p. 46). Underlying Ong's point is the fact that oral cultures, lacking any storage medium outside of human memory, practice economy in their time-binding, and only pass on what is needed for survival, only what is functional and useful. Historical and biographical details do not need to be preserved, especially if they are no longer relevant to the present. Oral societies are not bound by the weight of history in the way that literate societies are, and we might consider how well-balanced the world would be if all of its peoples would forget their historical conflicts, prejudices, and grievances.

Some time ago, a colleague told me that he was reading a book of *ancient* myths, and when I asked what myths they were, he said they were the myths of the Australian aborigines. I responded by noting that those myths could not be considered ancient, as they were only recorded a century or two ago. It is an easy enough mistake to make. I well remember watching documentaries when I was young that purported to be about a people "untouched by time," whose way of life was "unchanged since the dawn of time." But how can anyone know how much has changed if there are no historical records? What we do know is that homeostasis is not stasis, it is a dynamic equilibrium, evolving not in the progressive sense that we are accustomed to, not by accumulating increasingly greater amounts of knowledge, nor by making significant technological advancements, but simply by adapting only as much as is needed to maintain a balance in response to changing circumstances. As Ong (1982) puts it, oral cultures are "conservative or traditionalist" (p. 41), their main concern is to hold onto the knowledge that they already have, to maintain their precarious hold on whatever knowhow has been working for them, and they therefore tend to reject innovation and novelty, and venerate the wisdom of their elders.

Members of oral societies, therefore, live in the present, but they continually look to the past, and value the past. They typically talk about a mythic golden age that they long to return to, a time when the world was created, society was founded, a time of perfect unity and knowledge, like the Biblical story of the Garden of Eden. According to Mircea Eliade (1959, 1975), they find it relatively easy to move from the profane time and space of everyday life to a sacred time and space, one that connects directly to that moment or era of creation and foundation. The shift is accomplished through ritual, which is a dramatic reenactment of action that occurred during that golden age; oral performance becomes a ritual drama, and mimesis a form of communion. In this sense, oral cultures are certainly pre-scientific, and also pre-historical, and myth as the content of oral tradition is the functional equivalent of science and history. And my intent is not to suggest that there is something desirable about being pre-scientific and pre-historical, nor do I want to romanticize oral societies. But I do think it important to acknowledge that they *represent* an ideal of balance that we find both valuable and elusive, and that it is that characteristic of flexibility and homeostasis that Innis was hoping to see restored, rather than a wholesale return to tribalism.

What was it then, that pushed us out of balance? It was a complex set of factors that includes the agricultural revolution, the creation of stable settlements culminating in the formation of cities, the institution of complex social hierarchies leading to the establishment of a king and ruling class, the introduction of some form of economic system that accounts for trade and taxation, the appearance of some kind of organized religion that includes temples and priests, and more. But all of these different, disparate elements are all bound up with and bound together by systems of communication, specifically by systems of notation, and ultimately by the revolutionary innovation in communication that we call writing. Writing gave us a means to store knowledge outside of human memory, and Korzybski recognized that writing was a necessary prerequisite for a truly progressive form of time-binding. But writing also froze language in a relatively permanent form, replacing the flexibility of oral tradition with the rigidity of the fixed text. Homeostasis became harder to achieve after words began to be written in stone. And it was especially when writing was preserved by durable media such as stone, clay tablets, and the parchment codex, that the past stopped serving the present, and the present became the servant of the past. This unhealthy fixation with the past is what Innis (1951) meant when he wrote about time-biased cultures.

On this point, I differ with Carey (1989) and others who have ventured interpretations of Innis's dialectical approach, as I would argue that Innis did not intend to categorize homeostatic oral cultures as time-biased. Time bias implies a society that is unbalanced, that exhibits an unhealthy obsession with preserving the past and maintaining the status quo. And time bias implies a society that is dominated by some form of organized religion. The word *religion* itself is worthy of some attention, in that it is commonly said to have been derived from the Latin word for *binding*, implying a binding of human beings to the gods or God, a binding covenant expressed through ritual and dogma, a binding together of a congregation, and also, I think we can say, a binding of time. But no one is entirely sure of the origin of this word, and Cicero (1972) argued for a different derivation, one in which the root meaning of religion is *to reread*, *to read again*. Following Cicero's lead, I would suggest that tribal cults turn into organized religions when their rituals are written down, when the oral performance of ritual drama becomes a rereading of a written text, when the flexibility of ritual rooted in oral tradition becomes fixed in writing. And myth become religion when a changing repertoire of songs and stories featuring supernatural agents are written down and canonized as a sacred text, formalized and frozen, and preserved with great care, often guarded and controlled by a priestly class. Complex writing systems, such as cuneiform and hieroglyphics, and texts written in archaic or dead languages, help to enforce priestly *monopolies of knowledge*, to use the economic metaphor that Innis (1951) introduced. And control over texts in turn facilitates priestly control over sacred time, ending easy access to spiritual communion for the rest of the population.

Goody (1986) explains that the introduction of a sacred text transforms religious experience from a loose set of spiritual practices and beliefs, one that is fluid and flexible, to a set doctrine to which all must adhere. With the sacred text, a line is drawn between adherents who are members of the religious grouping and all the rest who are unbelievers and infidels; in other words, religion becomes an either/or affair, as in either you swear allegiance on and to the text and all that it contains, or you are an outsider, and if you are a member of one religion, you cannot be a member of another at the same time. With a sacred text, conversion becomes conceivable, and so does orthodoxy, fundamentalism, and heresy. Concrete images of the supernatural, in which the sacred is immanent, permeating the environment and surrounding us, give way to abstract conceptions in which the supernatural becomes distanced and transcendent, moving from the earth and water to a mountain top, from a mountain top to the sky, and from the sky to God knows where. As Innis notes, writing opens the door to monotheism, but even the polytheism of the Greeks and Romans becomes increasingly more abstract with literacy. With all this in mind, I would take the position that the term "religion" is best reserved for systems of belief and practice that are associated with writing, that the myths and rituals, and the cults and spirituality of oral cultures do not constitute the specialized institutions and coherent belief systems, bounded and binding, that we define as religion.

Time Bias vs. Space Bias

Given that the introduction of writing knocks cultures out of balance, innovations in writing technology can be seen as an attempt to restore that balance. One example would be the introduction of lightweight and transportable writing surfaces such as papyrus and paper, to offset the heavy media of stone, clay tablets, wood, and parchment. Such light media allow for a reliable means of sending messages back and forth over distances, and serve the administrative needs of the king and government, while also being useful for trade and commerce. Even more significantly, all media that facilitate communication over space are inherently military technologies, the contemporary phrase used for such functions being *command and control*. In this way, such new forms of writing allow for the growth of secular sectors of society, and make it possible for societies to expand beyond local territories, into kingdoms and empires. This then results in a new kind of imbalance, as the pendulum shifts to the other extreme, and we get the kind of culture that Innis (1951) referred to as space-biased.

Time remains an important consideration, however, but the need for preservation and durability is replaced by an interest in speed and transportation. Control also requires coordination and synchronization, which can best be achieved by systems of time-keeping and time-telling, such as the calendar in the ancient world, and the mechanical clock in medieval Europe (see Innis, 1951, 1972; Mumford, 1934). These technologies, which are based on writing and reading, break time down into homogenous units, years, days, hours, and as a consequence our experience of time changes. Edward T. Hall (1983) notes that traditional societies (e.g., oral cultures) are polychronic, that is members of such cultures see time as heterogeneous, continuous and unstructured in character, and they consequently treat time in a way that is flexible and open to what we call multitasking. Calendars and clocks move cultures in the direction of the monochronic, in which time is experienced as homogenous, uniform, and repeatable, linear and punctuated, so that punctuality is valued, and a focused, one-thing-at-a-time approach is common. Monochronic cultures reduce the experience of sacred time down to infrequent special occasions, holiday celebrations, while opening the door to the modern metaphor of time as money (Lakoff & Johnson, 1980). This ultimately leads to our contemporary notions of a 24/7, and 24/7/365 lifestyle.

Light and easy to use writing surfaces also facilitate copying, which not only undermines the time-bias of heavy media, but also restores some of the flexibility of oral tradition, since copying was rarely free from error, and scribes rarely concerned with exact duplication of documents. Another set of innovations that served to counter time biases were the simplifications of complex writing systems, such as the shift from cuneiform and hieroglyphics to phonetic writing systems, including the alphabet. This in turn led to the mechanization of writing through the invention of the printing press with moveable type, and the mass production and distribution of written works gave a great boost to the nascent space bias of Renaissance Europe. Ironically, however fragile and perishable each individual copy might be, the production and diffusion of multiple copies of the same text was more effective at preserving knowledge over time than the creation of a single copy in a highly durable medium, as Elizabeth Eisenstein (1979) makes clear. But the social impact was to undermine the time bias associated with the medieval manuscript, and break the monopoly of knowledge that the church held, which was based on its ownership and scribal copying of parchment manuscripts; the printing of works in contemporary vernaculars, rather than Latin and other learned languages, which printers did to increase their markets, further contributed to this process (Innis, 1951, 1972). All of these developments served to democratize writing and reading, and thereby disrupt pre-existing religious, political, and social hierarchies. This in turn led to the growth of scholarship, the critical examination of existing traditions, and the growth of knowledge. Robert Logan in his McLuhan-inspired work, *The Alphabet Effect* (2004), has shown how alphabetic writing in particular was intimately linked to the growth of science in western culture, through its particular ability to facilitate logical thinking, analysis, and classification; these effects take hold especially after alphabetic writing was amplified by typography.

The technologies of written communication, then, underlie both the conquest of nature and the conquest of peoples. Marxist critics have long noted the relationship between empiricism and imperialism, but were unable to explain the connection adequately—there is something more at work here than some conspiracy on the part of the bourgeoisie. We can understand the idea of progress in science and technology best by understanding that it is a spatial metaphor, the root meaning of *progress* being travel across territory. The very idea of progress over time originates as a by-product of a space-biased culture, and this amounts to a shift in time consciousness. Oral cultures look backwards to the past for legitimacy, for archetypes and models, and long to return to the moment of creation, a golden age, or at least recover the lost knowledge of their ancestors. But the introduction of writing, especially when coupled with a bias towards space gradually results in a turn away from the past and towards the future, as embodied in the idea of progress. The belief is that things are getting better over time, the present is superior to the past, and the best is yet to come (Perkinson, 1995). People *look forward* to the future, longing for tomorrow, whether it is imagined as a progression continuing on indefinitely, or as reaching an end state of utopia. The word *old* becomes a term of derision, and in the print era readers turn their attention to two new literary forms, the *novel*, and the *news*. Perhaps the conceptual shift is best summed up by the change in the meaning of the word *original*, which once only meant the first and oldest, coming from the moment of origin in the past, and has also come to mean the newest, most innovative, most cutting edge, or better yet, bleeding edge.

The spatial imbalance associated with the Egyptian, Alexandrian and Roman empires in the antiquity, later moved to the Mohammedan and Mongolian empires of the east, and then manifested in the commercialism, colonialism, and industrialism of modern Europe and America. Innis was profoundly concerned with the continued intensification of our space bias brought on by the application of electricity to communications, in the form of the telegraph, telephone, and broadcasting, which enabled us to engage in instantaneous communication over great distances. But he also held out some hope that an acoustic medium like radio might restore some semblance of the oral tradition, and thereby help to restore balance to western societies (see Innis 1951; see also Innis, 1952).

The Present Time

James Carey (1989) insisted that we can only understand people in the context of their particular time and place. And we can see in both Innis and Korzybski an attempt to respond to the terrible events of the twentieth century, which included the First World War, the Great Depression, the Second World War, the atom bomb, and the cold war, as well as the rise of propaganda and mass persuasion, including advertising and public relations, and the dominance of mass communication and mass culture. Can anyone blame them for hoping that it might be time, at last, for us to enter a new era of sanity and balance?

Of course, it is easy enough for us to say, some sixty years later, that Korzybski and Innis were wrong, that the kinds of changes that they envisioned never came to pass. But perhaps it would be more accurate to say that the changes did come to pass, only not in the way that they had hoped for. Korzybski's dream of a scientific society is not a reality, but Einstein's Theory of Relativity, along with Heisenberg's Uncertainty Principle are the cornerstones of what has been termed the postmodern condition, with its cultural and moral relativism. And while we ourselves may not be more rational today than we were a century ago, we live in a society guided by the rational principle of efficiency, the cornerstone of what Neil Postman (1992) referred to as technopoly, the surrender of culture to technology. And we turn increasingly greater portions of our affairs over to that supreme engine of mathematical action, the computer. Where Korzybski wanted us to be better human beings, we have instead been taking the human element out of the equation, and automating the process. Rather than entering a mature phase, an adulthood of humanity, we seem to find ourselves in a new form of infancy.

And despite Innis's hopes, oral traditions seem more distant than ever before in the age of television and the internet. We have experienced a continued growth in sonic technologies, however, which Ong (1967, 1982) termed secondary orality to emphasize their distinction from the primary orality of oral cultures. McLuhan (2003) spoke about a retrieval of acoustic space, but that amounts to an embrace of nonlinearity and subjectivity, rather than a retrieval of memory and dialogue. McLuhan also talked about retribalization and the global village, but again this represents something quite different from preliterate tribalism and village life, as it involves instantaneous global telecommunications. We have found a new kind of interactivity made possible by computer-mediated communication, social networking, and social media, and this does seem to provide us with a form of communication that resembles orality in certain respects. But are a series of updates and comments on Facebook, MySpace, and Twitter the equivalent of oral dialogue? Does blogging take the place of epic poetry and public address? Can online groups and bulletin boards replace communities where individuals must cooperate out of necessity, in response to the requirements of material reality? Does the ephemeral nature of electronic communications, with websites and people's profiles vanishing overnight, provide us with the continuity that we so desperately need?

In one sense, electronic surveillance, and data collection and storage, present us with the possibility of balancing the space bias of western societies with a new form of time-binding, one so thorough and complete that it has been dubbed *total recall* (Strate, 2003). Does this go so far as to threaten us with a return to a time-biased way of life, one that would support and encourage the various fundamentalist and theocratic movements in existence today? Digital databases are easy enough to alter, it is important to note, and such alterations can be difficult if not impossible to detect. In this way, digitality does restore some of the flexibility of orality, and perhaps offer some promise for restoring homeostasis. Would Innis be encouraged? I suspect not, because contemporary digital alterations are not kept in check by a conservative or traditionalist worldview, and therefore are open to relentless revisionism, a kind of temporal anarchy. If there is potential for homeostasis here, it is a dystopian balance where again we find that the human element has been removed. The flexibility of oral tradition is based on the medium of human memory, the basis of human knowledge, for as Ong (1982) reminds us, "you know what you can recall" (p. 33), and as Fraser (1987) argues, computers do not really remember anything.

Where oral cultures naturally look to the past, and literate cultures have the potential to turn around and look towards the future, our electronic culture seems to be fixated on the present; the instantaneity of telecommunications communicates to us in the present tense (Strate, 2003). Even when the content is a recording or film, the broadcast signal creates the message in the present, and there is always the possibility of someone interrupting the broadcast to *bring us a special message*. We are plugged in, tuned in, our nervous systems "extended in a global embrace," as McLuhan (2003, p. 5) puts it. We are consequently impatient, intolerant of delays of even the slightest measure, as we live in what Jeremy Rifkin refers to as a nanosecond culture (Rifkin, 1987). With electronic updates transmitted online and onto mobile devices, it is not surprising to hear students say: *You mean you want me to pay to read on paper what I can get for free on the internet, and you want me to pay for yesterday's news?* We thrive on the live, the up-to-the-minute, the on-demand, the just-in-time. And our popular culture, popular psychotherapies, and popular spiritualities constantly advise us to *live in the moment*. While there is some utility to this advice, it is repeated over and over as if it is some kind of cosmic revelation, rather than a widely shared common sense assumption that is never called into question anymore. *Carpe diem! Seize the day!* Or so says Robin Williams in the 1989 film, *The Dead Poets Society*, which is presented to us as a model of what schooling ought to be like, contradicting centuries of our best time-binding efforts.

Our present-centeredness is more than a matter of the immediacy of electronic transmission and being online all the time, however, as we have also sought to bring the past and the future under the control of the present. Our sense of the past has previously been governed by narrative, whether it was the episodic storytelling associated with oral myth, or the linear accounts of written history. With digitality, the database substitutes for the narrative (Manovitch, 2001), and we are left with a discontinuous set of images, audio and video clips, nostalgic anecdotes, and fragmented historical documents, archival materials, and museum pieces, all randomly accessible, but lacking in coherent order, context or explanations, let alone detailed chronology. In giving us total recall, the digital database produces information overload, and in giving us a highly flexible form of time-binding, the digital database turns historical fact into a matter of personal opinion. We can individually construct the past as we see fit, just as we can construct our own personal newsfeeds and readers and newspapers on the web, and our own playlists for digital music, and just as we construct our own story out of the multitude of possibilities in a hypertext and a computer game.

We have also blurred the once-clear distinction between a performance and its recording through the use of computer programming. A program does not play back a performance, it is itself a performer, producing an automated performance (Jones, 1992). Each performance is a new "live" performance, but one that was constructed in the past, and each performance is identical to every previous performance, and every performance that will be repeated afterwards. The programmed performance brings the past into the present not as a recorded artifact, but as an event newly recreated in each iteration. This not only brings the past but also the future into the present. The program is an attempt to colonize the future on the part of the present (Strate, 2003). Programming the future should not be confused with planning for the future, which is what we did when we were forward-looking. Planning involves contingencies and uncertainties, hence Robert Burns famous lines from his poem, "To a Mouse," that "the best laid schemes o' mice an' men, Gang aft agley," and hence the Yiddish proverb, "mentsch tracht, Gott lacht," which is typically translated as "man plans, and God laughs," not to mention John Lennon's poignant lyrics from the album he released shortly before his assassination, "life is what happens to you while you're busy making other plans." Without a doubt, our children know the difference between making plans to play after school, and the kinds of programmed afternoon activities that they often are involved in. Programming is not so much about progress as it is about controlling the future, not so much about continuity as it is about uniformity and eliminating uncertainty. While the Long Now Foundation's goals are commendable, their choice of name is symptomatic of our contemporary form of time-binding. We live in a *long now* that extends far into our past and that we are trying to extend far into our future. But the problem with programming the future, as opposed to planning for it, is that programming is an attempt to eliminate human judgment, to bring the future into the present by means of hyperrationality,

93

as we bring the past into the present by means of hyperreality.

The very term *postmodern* reflects our inability to come to terms with the future as future, which is entirely dependent on an understanding of the past as history, and as *prologue*, as *the words that have come before*. *Postmodern* is not a word of explanation, it is an expression of the temporal confusion brought on by our present-centeredness. This same temporal confusion is reflected in the *Terminator* movies and television series, with its time-loop paradox that defies logical analysis. It is reflected as well in movies such as *Memento*, and generally in all the narratives where memory, identity, and reality are called into question, including the *Matrix* and its sequels, *The Bourne Identity* and its sequels, *Pleasantville*, *The Truman Show*, *A Beautiful Mind*, *Dark City*, *Blade Runner*, and of course, *Total Recall*. The list is quite extensive, including recent television series such as *Life on Mars*, *Journeyman*, *FlashForward*, and *Lost*, as well as movies where the future and the present blur together like *Minority Report*, and *Premonition*. Although it can be argued that we have created a new kind of sacred time, a present in which all times past and future intersect, and while that experience may exist for some individuals, in a more telling sense, we have created a completely profane time, 24/7/365, and entirely uniform. The distinction between sacred and profane time is dependent on maintaining some vestige of polychronic culture, however, and cannot be sustained in our monochronic society of the long now. In losing our sacred time, we are losing our much-needed Sabbaths, losing our opportunities for rest and reflection. And so we find that we are losing the distinctions between night and day, between the seasons, and that we are losing our past and our future. If time is a form of energy, then time itself is subject to time's arrow, is subject to entropy, the loss of quality, and this loss of quality time amounts to a loss of differentiation of time, a descent into a maelstrom of temporal chaos; rather than the heat death of the universe, it is the time death of humanity.

Having said all that, I do not believe that all is lost. Although Korzybski's dream of training people to think and act with enhanced rationality never quite materialized, I do think we have seen great success in the effort to combat the irrationality of stereotyping and prejudice, an area where Korzybski's program of general semantics has made significant contributions. And while we have yet to achieve the flexibility and balance that Innis valued, we have become more concerned with homeostasis, more ecologically minded, in many ways, especially in regard to the natural environment. There is no question that we still need to make much more progress in these areas, but following the advice of Wendell Johnson (1946), we also need to recognize and celebrate the progress that we have achieved.

Korzybski and Innis represent different concepts of time, different positions on how human beings ought to relate to time, but they are in many ways quite complementary. Korzybski valued progress, and I have argued that we need to retrieve and reclaim that word, and stop feeling embarrassed about using it. We have made great progress, and not just in science and technology, and we ought to feel good about how we have made things better, at least in certain ways. But we have to bring back the idea of progress in the holistic sense that Korzybski asked for, not as applying to specialized sectors of society relating to science and technology. We have to insist that it can only be called progress if it includes social, political, and economic progress, and moral, ethical, and ecological progress. At the same time that we need to move forward, we need to regain and then maintain our balance. We need a balance between progress and continuity, between the individual and the community, between the profane and the sacred, between science and religion, between technology and ecology, between space and time. We need to put an end to the tyranny of the now, which means that we have to actively counter the biases of our contemporary electronic media environment, following what Postman (1979) argued was the thermostatic function that schools ought to be carrying out in order to help us find a measure of homeostasis.

That means that we need to teach history as a coherent narrative, or set of narratives, narratives that help to contextualize the present, that show the progress and the backtracking, the discoveries and the errors, the good and the evil, so that we can understand ourselves, as a species, in time; and this includes the history of communication, and the arts, the religions, the philosophies, and the sciences and technologies. And we need to teach the history of the future, and the future of the future, futurism not as being about entrepreneurial efforts and the introduction of new products, but about planning and conserving, about preserving and preparing for the generations to come, about achieving and maintaining sustainability, about pondering the impact and effects of innovations, and the fact that change is always unpredictable and needs to be approached with great care. The past and the future need to be in balance with one another, with the present serving as an appropriate fulcrum between the two. And we need to make our time balance on a human scale.

      We exist only because we are riding on that Big Bang that happened some fourteen billion years ago. We are alive because we are riding on a second big bang that occurred about four billion years ago on this planet, the origin of life. And we are here to talk about it because we are riding on a third big bang that occurred maybe forty thousand years ago or so, the origin of language. As a species, we are binders of time, bound up by our biases of time; we are moved by our consciousness of time, as we tell time, and as we tell ourselves that only time will tell; as we play for time, and as we pray, as we pray for time.

Chapter Five

Post(modern)man

H aving been a student of Neil Postman's, I would like to affirm that teaching is a *Postman* activity, and I want to acknowledge the debt that I and many others owe to him as an educator. I would add that in the many years that I have known him, very little of what I have heard him say might be categorized as crazy talk or stupid talk.[*] A good portion of it, however, has been amusing. For even in the midst of the most conscientious of objections, Neil Postman never loses his sense of playfulness. I am not sure why this is so. Perhaps it is because he is the youngest of four children. Perhaps it is because he is from Brooklyn. Or maybe it has something to do with his name. After all, how many people do you know whose first and last name constitute a complete grammatical sentence (*Kneel, postman!*)? If naming is destiny, then his surname also may account for his interest in communication, as "Postman" summons images of messengers, messages, and media, particularly of the pre-electronic variety. Moreover, given our current era of epilogues, this age of *post*structuralism, *post*Marxism, *post*feminism, and of course, *post*modernism, the name *Post*man seems especially timely. But whatever the relationship between the word and the thing it represents, this much is clear: His is a name that is rich in meaning, a name that invites wordplay. And it is in this spirit that I have taken his name and conflated it with the term "postmodern" in order to arrive at the title of this paper, "Post(modern)man." In doing so, I wish to suggest that we can gain some insight into Postman's perspective on media and technology by framing it as a theory of the postmodern. I also mean to imply that those who are interested in the concept of postmodernism would benefit from a review of Postman's scholarship.

I should make it clear at this point that Neil Postman has never claimed to be a postmodernist.[**] Rather, he has referred to himself as an educationist, that being his original area of expertise (see Postman, 1961, 1979, 1988, 1995; Postman & Weingartner, 1966, 1969, 1971, 1973); as a general semanticist, having served as the editor of *ETC* for many years

---

[*] This essay was written a decade before Neil Postman passed away, and therefore refers to Postman in the present rather than past tense. I have elected not to modify it, leaving it in its original form for sentimental reasons.

[**] Indeed, he expressed his disapproval of the characterization in his last major work, *Building a Bridge to the Eighteenth Century* (Postman, 1999).

(also see Postman, 1976, 1988, 1995; Postman & Weingartner, 1966; Postman, Weingartner, & Moran, 1969); and as a media ecologist (see Postman, 1979, 1982, 1985, 1988, 1992; Postman, Nystrom, Strate, & Weingartner, 1987; Postman & Powers, 1992, for his scholarship on media and technology) . Media ecology, a phrase inspired by Marshall McLuhan, is the name Postman gave to the graduate program that he founded at New York University, and the term that he has used to refer to the general perspective of McLuhan (1962, 2003) and others such as Harold Innis (1951), Eric Havelock (1963, 1976, 1978, 1982), Walter Ong (1967, 1977, 1982), Lewis Mumford (1934), and Jacques Ellul (1964, 1965, 1985). I should also make it clear that these scholars, in turn, have never claimed to be media ecologists, the point being that self-identification is irrelevant. Still, it should be duly noted that not only has Neil Postman never referred to himself as a postmodernist, but he never, to my knowledge, uses the term "postmodern.[*] Nor does he use any of the jargon associated with postmodernism, such as "decentering," "hyperreality," or "pastiche." In short, Postman does not speak postmodern. He speaks English. I would argue, however, that it is not necessary to speak postmodern in order to speak *of* the postmodern, and that Postman's clarity could serve as a welcome corrective to the excessive use of jargon and notoriously esoteric writing that is typical of postmodernists (see, for example, Baudrillard, 1981, 1983, 1988; Jameson, 1991; Lyotard, 1984).

The difference in linguistic style is, at least in part, symptomatic of differences in intellectual background. Postman is very much a part of the Anglo-American tradition of empiricism, utilitarianism, and especially, pragmatism, particularly as manifested in the fields of education and communication; among his chief influences are John Dewey, Karl Popper, George Herbert Mead, Alfred Korzybski, I.A. Richards, and of course Marshall McLuhan. Postmodernism, on the other hand, is firmly rooted in continental philosophy, in Nietzsche, Marx, and Saussure, and is intimately intertwined with the disciplines of art and literary criticism. There is some common ground, however, as postmodernists such as Jean Baudrillard (1981, 1983, 1988) and Frederic Jameson (1991) also list McLuhan as a major influence; their path to McLuhan was simply somewhat longer and more convoluted than Postman's.

---

[*] Except, again, in the brief critique he included in his final work (Postman, 1999).

Diagnosing the Postmodern Condition

Having gone on at some length about postmodernism and postmodernists, I realize that some discussion of the concept of the postmodern itself is overdue. In its most basic sense, the term "postmodern" implies that a period of time that has been labeled "modern" has ended, and that we find ourselves in new and uncharted territory. There is, of course, a certain irony to the term "postmodern" if we interpret it as meaning "post-contemporary" (a phrase actually used by Jameson, 1991), but this is ultimately a reification of the term "modern". For rather than referring to our present point in time, "modern" here signifies a historical era that comes to a close some time during the mid to late twentieth century. When this time period is said to begin is subject to some variation, however. Critics who focus on the history of the arts, and on the modern as an aesthetic style, place the beginning of the modern period at the turn of the twentieth century. Clearly, this concept of the modern is relatively narrow and specialized. Social theorists, on the other hand, trace the origins of the modern back to the late eighteenth and early nineteenth centuries, and to such factors as the rise of democracy, capitalism, and urbanization, and the Enlightenment, the industrial revolution, and mass culture (Lyotard, 1984; Jameson, 1991; see also Ewen, 1988, for a familiar but comprehensive summary of these historical developments).

There is no doubt that these developments are of great importance, but they are overshadowed by the periodization used by Postman (1979, 1982, 1985), inspired by McLuhan (1962, 2003), and perhaps expressed most clearly in the work of Elizabeth Eisenstein (1979); according to their chronology, the nineteenth century may have seen the beginnings of a *late* modernity, but this follows an *early* modern period that begins in the fifteenth century with the European printing revolution. Postman, McLuhan, and Eisenstein argue that the changes brought on by this revolution result in the termination of the medieval period and set the stage for the further mutations that occur during the eighteenth century. In short, they argue for the essential unity of this larger modern period, a period also known as the "age of Gutenberg" (a phrase used by Postman as early as 1961, one year prior to the publication of McLuhan's

Gutenberg Galaxy). Thus, from Postman's perspective, postmodernity represents the end of five hundred years of print-dominated culture, a state of affairs that he laments.

In this context, it is worthwhile to note that postmodernists do not necessarily celebrate the postmodern; many are quite critical of it (e.g., Baudrillard, 1981, 1983, 1988; see also the discussion in Jameson, 1991, chap.2). And this brings me back to the title of this essay, "Post(modern) man," and to a second meaning that it holds for me: that Postman *contains* the modern, that he acts as a champion of the modern in the postmodern world. In making this assertion, I realize that I am placing myself at odds with Joli Jensen who argues in *Redeeming Modernity* (1990) that Postman is hostile to the modern; her use of the term "modernity" however does not clearly distinguish between the modern as a historical period with distinct boundaries and the modern as simply the contemporary. Instead, I would suggest that Postman is a defender of modernity, and in particular of print culture as the better part of modernity. He is at his best when he gives voice to print culture, acts as an avatar of typographic discourse, and plays the role of "Minerva's owl" in the "gathering dusk" of postmodernity (see Innis, 1951, p.3, for a discussion of this quotation from Hegel).

In keeping with this role, Postman not only sounds the alarm, but also identifies those forces responsible for the assault on modernity. They include, of course, television. Like McLuhan (2003), Postman links the adoption of television technology to the demise of the modern. He differs from McLuhan, however, in his explanation of this phenomenon. McLuhan focuses on the relationship between technology and the senses, arguing that after five hundred years of eye dominance through typography, television has restored the ear to its previously held position of superiority. Walter Ong (1967, 1977, 1982) echoes McLuhan in referring to our age as one characterized by "secondary orality." Postman, on the other hand, is concerned with the relationship between technology and discourse. Consequently, while he acknowledges the distinction between orality and literacy, he focuses on what is common to all forms of language; he is a defender of the word, not just the printed word, but also the handwritten word and the spoken word. For him, the eloquence of print culture is rooted in a balance between what is read and what is said. This balance has now been upset by televisual discourse, which shifts the emphasis from verbal to *visual* forms of communication. In other words, Postman's concept of postmodernity is that of a culture in which the *image* has come to overshadow the word (in this respect he is in perfect agreement with other postmodernists such as Baudrillard, 1981, 1983, 1988, and Jameson, 1991). While other postmodernists become preoccupied with hyperreality, simulation, and the relationship between signifier and signified, Postman focuses on the concreteness of the image in comparison with the word's capacity for abstraction, on the inability of pictures to represent propositional statements, on the association between visual communication and pathos as opposed to verbal communication and logos, on the relative accessibility of iconic forms in contrast to the long process of schooling associated with literacy, and, of course, on the image's tendency to amuse, entertain, and ultimately, to trivialize.

The shift from linguistic to image-based discourse is a key element of Postman's perspective on postmodernity, but he uncovers other aspects of televisual discourse that are hostile to the modern. For example, the continuous, linear, and logical arguments favored by the moderns are left behind in the dust of instantaneous electronic communication's accelerated discourse (for more on the relationship between speed and postmodernity, see Virilio, 1986; Virilio & Lotringer, 1983); ultimately, it is a form of discourse that is present-centered and therefore ahistorical (see Jameson, 1991, for the connection between postmodernism and the loss of a sense of history). Also, telecommunications all but eliminates the concepts of distance and location that the moderns went to such great effort to map out, resulting in a discourse that is decontextualized and often of little direct relevance to postmodern populations (in this respect, McLuhan's, 1962, 2003, concept of the global village is well known, but see also Jameson's, 1991, emphasis on the connection between postmodernity and the rise of multinational capitalism). Moreover, our new communication technologies have dramatically increased the volume of information transmitted in our culture, crippling the notions of secrecy and privacy that were constructed through typographic discourse, resulting in information overload and the inability to process and control information (also see Lyotard's, 1984, arguments concerning the relationship between science, information technology, and the disappearance of metanarratives).

Postman's emphasis on television might seem to imply that, like many other postmodernists, he views the end of modernity as a sudden collapse and surrender. On the contrary, he traces the attack on modernity back to the nineteenth century's revolution in communication technologies, such as photography and telegraphy. The evolution of audiovisual media such as the motion picture and radio put modernity on the retreat. The art world's discovery of the modern and its transformation into an aesthetic style at the turn of the twentieth century was, as McLuhan would conclude, a sign of modernity's obsolescence. Thus, the fall of modernity associated with the widespread adoption of television during the 1950s and 1960s was part of a longer process of decline. This more gradual view of the advent of postmodernity is consistent with Postman's use of the extended modern period associated with the printing press; modernity itself develops gradually from the invention of typography through the incunabula of the early modern era.

Postman's arguments about the triumph of the televisual over typographic discourse are the best known elements of his perspective on postmodernity, but they do not represent that perspective in its entirety. Rather, he also points to the triumph of the technological over the traditional (see Postman, 1992, and also 1976, pp. 178-185, and 1979, chap. 5). The modern period, he argues, gave rise to a technocratic culture in which traditional values and customs coexist with an emerging scientific and technological worldview. In this case, the assault on modernity occurs when this balance becomes upset, and the technological, with its emphasis on efficiency, takes command. (This coincides with Jameson's, 1991, description of the modern era as a period marked by the process of modernizing, whereas postmodernity is a sign of a state of full modernization). Postman refers to this aspect of postmodernity as technopoly, a culture monopolized by the technological. Here again, the shift is not a sudden one, as Postman traces the roots of technopoly to the early twentieth century, and the development of scientific management; full blown technopoly seems to be associated with the post-war development of computing technology and the creation of what is sometimes known as the information society. He argues that it is a society characterized by information overload and an inability to screen out or evaluate information, except by technology's own criteria of progress and efficiency. (This corresponds to Lyotard's, 1984, description of postmodernity as characterized by the absence of any ruling ideas, narratives, and myths,[*] and to Jameson's, 1991, explanation of postmodernism as the cultural logic of late capitalism, of a culture in which capitalism has penetrated to all sectors of life; technopoly and late capitalism may be viewed as competing explanations of the same phenomenon, and/or two sides of the same coin.)

Mediating the Postmodern Contradiction

Thus, Postman presents us with two paths to postmodernity: from print media to electronic media, and from technocracy to technopoly. What he has yet to do[**] is to explain in its entirety the relationship between these two paths. There is no doubt that a connection exists between typographic discourse and technocracy; Eisenstein (1979) has amply documented that printing technology was a necessary precondition for the development of modern science and technology. The association between the computer and technopoly is also quite clear. What is something of a puzzle is the relationship between technopoly and television. Of course, technopoly's preeminence insures that television technologies will be accepted and adopted without doubt or question. Television in turn serves as technopoly's user-friendly interface; in its programming as well as its advertising, it functions as the great communicator and promoter of the technological. And yet, there is a fundamental contradiction between televisual discourse and technopolistic discourse. Whereas televisual discourse trivializes what is truly important, technopolistic discourse gives the truly trivial the illusion of importance. Whereas television amuses us to death, technopoly's information glut gives us all anxiety attacks. And whereas the discourse characteristic of television might be characterized as irrational, that of technopoly may best be described as hyperrational. The postmodern, therefore, is a product of these two opposing forces; it is a culture pulled in two directions at once, a culture that is, perhaps, strained to its limits and in danger of being ripped apart. The question, then, is can Postman's perspective account for the cultural contradictions

---

[*] Postman's own version of this argument was presented in his penultimate work, *The End of Education* (1995).

[**] And never got around to doing, as his focus was on addressing issues of the day, not theory-building or the construction of a philosophical system.

of postmodernity?

I believe the answer is yes, and I will extrapolate from Postman's scholarship in order to provide an explanation. As I have noted, Postman favors modernity for its high level of verbal discourse, for its emphasis on the spoken, written, and printed word. Televisual discourse, on the other hand, is largely based on the image, and it is this iconic discourse that is seen as responsible for "the humiliation of the word," as Ellul (1985) puts it. But I believe it fair to say that the word would never have lost to the image had the fight been a fair one, that, in fact, the word was outnumbered, and forced to do battle on two fronts. The other enemy of the word, the image's ally in its victory of the postmodern, is none other than the number. In other words, technopolistic discourse is quantitative discourse, so that it is not so much any particular type or class of technology that is at the core of technopoly, but rather it is the worldview which not only believes but demands that all reality be reduced to numeric form, measured and statistically manipulated in order to maximize efficiency. In sum, the postmodern world is one in which the image and number have outflanked verbal communication and have emerged victorious in the battle for control of our culture and our collective consciousness.

While the biases of the visual and the numeric contradict each other, it should be noted that the two codes have had a long and distinguished relationship, a relationship based on developments such as Euclidean geometry and Cartesian coordinate graphing. In many ways, television reflects the uneasy tension between these two forces, for while its content is dominated by the image, the television industry is run almost entirely by the numbers, by audience ratings, shares, and demographics, and by profits (Postman, 1988; Postman et al, 1987; Postman & Powers, 1992). And while the computer has long been viewed as a number-cruncher, widespread adoption of computer technology has been dependent on the development of an iconic graphical user interface (aka GUI, e.g., that of Macintosh or Windows). Moreover, the visual and the numeric have come together in the form of the digitized image, which represents digital technology in the service of analogic communication. I would suggest to you that the digitized image, free of any necessary relationship with reality, but rather in the realm of pure simulation, represents the ultimate expression of postmodernity's triumph over typographic discourse and the epistemology of the printed word.

This brings me to a third meaning for my title, "Post(modern)man": that it refers to the type of "man" or rather the type of person characteristic of postmodern culture. In his book *The Disappearance of Childhood*, Postman (1982) explores one aspect of the postmodern population, that it is one in which the distinction between childhood and adulthood has all but disappeared (this is an example of what postmodernists call the decentering of the subject). Postman is not content simply to describe the postmodern sense of self, however. Instead, he has an idea of how postmoderns might be shaped, of how the negative effects of postmodernity might be mitigated. His answer, the answer that runs through just about every one of his books, is education. In his role as Minerva's owl, it is the school, more than any other of typographic culture's institutions, that Postman seeks to salvage. It is schooling, education in the book and by the book, that he offers as a mediator between the binary oppositions of image and number. Postman is a postmodernist because he does not deny that modernity has passed. But he is more than a postmodernist because he does not just accept and describe postmodern culture, but rather seeks an alternative in the form of education. In doing so, he reminds us that teaching is, after all, an activity, not a form of passivity.

Chapter Six
Defender of the Word

N eil Postman[1931-2003] died on Sunday, October 5th, 2003, at the age of 72, after battling lung cancer for almost two years. His contributions, as a scholar, teacher, and public intellectual, enriched many different fields of study, including semantics, linguistics, communication, media studies, journalism, education, psychology, English, cultural studies, philosophy, history, sociology, political science, religious studies, and technology studies, etc. Across these many contexts, and throughout his career, he promoted and advanced the discipline of general semantics. Some years ago, in the course of a conversation we shared on the writing styles that intellectuals and academics employ, he summed up his position on language with these words: "Clarity is courage." Postman wrote and spoke with a crystalline courage, and championed clarity in language, thought, and action.

Born into a Jewish family in Brooklyn, New York, exposed to Yiddish and Hebrew in addition to English, Neil Postman[1931] developed an awareness of the power of language at an early age. Public school education at that time placed a great deal of emphasis on proper grammar and diction, and the elimination of accents and dialects. Consequently he learned to speak in the educated New York manner and idiom made famous by Franklin Delano Roosevelt.

Neil Postman[1950s] established himself as a star athlete on the varsity basketball team at the State University of New York at Freedonia, played minor league baseball, served in the United States Army, and studied for his doctorate in education at Columbia University's Teachers College. His mentor, Louis Forsdale, introduced him to the formal study of linguistics, to the fields of education and communication, and to the study of media. Forsdale also introduced Postman to Marshall McLuhan, the University of Toronto English Professor who would become famous during the 1960s for his study of media.

Neil Postman[1958] joined the English faculty at San Francisco State College, where he shared an office with Mark Harris, author of *Bang the Drum Slowly*, and worked under S. I. Hayakawa. Largely through Hayakawa, Postman learned about general semantics, and became a member of the International Society for General Semantics. As a doctoral student at Teachers College, Postman had studied linguistics, and New York University hired Neil Postman[1959] for its School of Education on account of his expertise in that field. General semantics fit neatly within his linguistics orientation, and he inherited possibly the first college course in the subject at New York University. In Postman's (1988) own words:

> I have been unable to verify the exact date but there is suggestive evidence that in the late 1940s, NYU's School of Continuing Education sponsored a seminar given by Korzybsi himself. And in Stuart Chase's popular *The Power of Words,* Chase asserts that an NYU School of Education course called "Language and Behavior" was among the first general semantics courses ever given at a major university. That course survives to this day under the title "Language and Human Behavior." (p. 145)

In teaching "Language and Human Behavior" over four decades, Postman made it into the oldest continuously taught course on general semantics in the history of the discipline.

Neil Postman[1960s] focused on English education, arguing that we could improve the English curriculum in elementary and secondary schools by incorporating linguistics and semantics, as well as the study of "the new languages," a phrase that McLuhan's colleague Edmund Carpenter (1960) introduced to refer to the media of communication. Thus, Postman's first book, *Television and the Teaching of English,* commissioned by the National Council of Teachers of English (through Forsdale), and published in 1961, clearly indicated the direction his career would take. He then went on to develop a textbook series called "The New English," used in grades 7 to 12. Published between 1963 and 1967 under the titles *Discovering Your Language* (Postman, Morine, & Morine, 1963), *The Uses of Language* (Postman & Damon, 1965a), *Exploring Your Language* (Postman, 1966), *The Languages of Discovery* (Postman & Damon, 1965b), *Language and Systems* (Postman & Damon, 1965c), and *Language and Reality* (Postman, 1967), this highly innovative series became quite popular in classrooms across the United States. Through these books, Postman introduced a generation of students to the principles of general semantics and related perspectives on language and symbolic communication. Postman's first collaborative effort with fellow Teachers College graduate Charles Weingartner gave us the theoretical context behind his textbook series. Entitled *Linguistics: A Revolution in Teaching,* and published in 1966, the book presented a scholarly synthesis especially for the benefit of English teachers.

Postman's calls for a new approach to English education fit together with the growing movement for educational reform during the 1960s, and the publication of *Teaching as a Subversive Activity* (coauthored with Weingartner) in 1969 catapulted Postman into a leadership position in the movement.   Combining linguistics, general semantics, and McLuhan's ideas about media, and criticizing the American educational system in general this time, rather than just the teaching of English, Postman and Weingartner called for a curriculum based on the "Sapir-Whorf-Korzybski-Ames-Einstein-Heisenberg-Wittgenstein-McLuhan-Et Al. Hypothesis . . . that language is not merely a vehicle of expression, it is also the driver; and that what we perceive, and therefore can learn, is a function of our languaging processes" (p. 101). In this new model of education, understanding language (including the new languages of media) would take the leading role. Teachers would emphasize the art of asking questions and what Postman and Weingartner called "the inquiry method" (p. 25), and the evaluation of statements or as they put it, "crap detecting" (p. 1). *Teaching as a Subversive Activity* had a dramatic impact on the reform movement, and remains influential to this day. And Postman and Weingartner produced two additional books on education, *The Soft Revolution: A Student Handbook for Turning Schools Around* in 1971, and *The School Book: For People Who Want to Know What All the Hollering is About* in 1973. They also co-edited, together with Terence P. Moran, the anthology, *Language in America*, published in 1969; Postman contributed a chapter on the misuse of language entitled "Demeaning of Meaning." In this chapter, Postman employs terms such as "semantic environment" and "language pollution," in presenting an argument that would become central to his media criticism of the 1980s:

> In considering the ecology of the semantic environment, we must take into account what is called the communications revolution.   The invention of new and various media of communication has given a voice and an audience to many people whose opinions would otherwise not have be solicited, and who, in fact, have little if anything to contribute to public issues. Many of these people are entertainers, such as Johnny Carson, Hugh Downs, Joey Bishop, David Susskind, Ronald Reagan, Barbara

Walters, and Joe Gargiola. Before the communications revolution, their public utterances would have been limited exclusively to sentences composed by more knowledgeable people, or they would have had no opportunity to make public utterances at all. Things being what they are, the press and air waves are filled with the featured and prime-time sentences of people who are in no position to render informed judgments on what they are talking about: like Joey Bishop on the sociological implications of drugs, Johnny Carson on education innovation, Ronald Reagan on the *Pueblo* incident, David Susskind on anything, and Hugh Downs on menopause. ("It is," he says, "a controversial subject.") (p. 14)

As this passage implies, "the ecology of the semantic environment" refers to essentially the same idea as the concept of media ecology, which incorporated the study of symbols, symbol systems, and symbolic form with the study of media and technology. Postman formally introduced the term "media ecology" in 1968, in an address to the annual meeting of the National Council of Teachers of English (publishing a revised version in 1970 as a book chapter entitled "The Reformed English Curriculum"). He also prepared a proposal for a new graduate program in media ecology at New York University, which he incorporated into *The Soft Revolution*. Approved in 1970, the media ecology curriculum included major works in general semantics, linguistics, and the philosophy of symbolic form as required reading, along with the scholarship of Norbert Wiener, Edward T. Hall, Erving Goffman, Paul Watzlawick, and of course Marshall McLuhan, Eric Havelock, Lewis Mumford, etc.

The combination of general semantics and communication theory that served as the foundation of the media ecology curriculum also became the basis of Postman's 1977 book, *Crazy Talk, Stupid Talk*. Popular among general readers as well as communication educators, *Crazy Talk, Stupid Talk* constitutes an in-depth examination of the semantic environment. That same year, Postman followed in Hayakawa's footsteps and became the editor of *ETC*, a position he held until 1986. As editor he sought to advance the discipline of general semantics by broadening its scope. He therefore expanded the focus of the journal to incorporate media ecology, and published a great deal of seminal work in the field, along with more traditional analyses and discussions in the tradition of Korzybski, Hayakawa, and Johnson. Thus, in his Keynote Address at the 1980 International Conference on General Semantics in Toronto, he tied together the symbolic analysis of Korzybski with the media analysis of McLuhan, Havelock, Carpenter, Innis, and other members of the Toronto School of media ecology:

> This focus on the structure of technics is the arena of inquiry staked out by Innis, who believed that embedded in every medium of communication is a bias toward either time or space. This is the arena in which Innis' most well-known student, McLuhan, has made his mark by probing the extent to which each medium amplifies or deadens one or more of our senses. And, at the risk of offending, I submit that were Korzybski alive today, he would be at the forefront of research in this same arena, for he taught that any medium which conveys a message carries in its structure and mode of presentation messages of its own. He understood better than anyone else that a medium is not a neutral mechanism through which a culture conducts it business. It is by its very form a shaper of values, a masseuse of the senses, an advocate of ideologies, an exacting organizer of social patterns. Korzybski, of course, focused his attention on the medium of language, but how fascinated he would be by the various forms of human communication we must cope with today. (Postman, 1980, pp. 322-323)

Postman's marriage of media ecology to general semantics came across as a welcome innovation to some, and a shotgun wedding to others, but overall an adequate assessment of this period in *ETC*'s history does not yet exist. However, Thom Gencarelli, (2000, 2006) in his research on Postman and media ecology, suggests that Postman's editorial work on *ETC* constitutes a major turning point in his career (see also Lum, 2006). During this period, Postman emerged as a leading media critic and public intellectual, much like Marshall McLuhan had become during the sixties.

Neil Postman[1979] published *Teaching as a Conserving Activity*, and in doing so explained that he had had a change of heart, distancing himself from Neil Postman[1969] and his earlier positions on educational reform. Whereas Postman[1969] had concluded that schools needed to change in order to adjust themselves to the new cultural environment dominated by television and the electronic media, Postman[1979] came to the realization that young people do not need any help in adjusting to television, but rather needed the print-oriented counter-environment that traditional schooling provided. He identified television as a curriculum in its own right, one that rivaled that of the traditional school, one in which images overshadowed words, and capturing attention overruled coherence. However much this amounted to a reversal of the position he took in *Teaching as a Subversive Activity*, Postman remained constant in his insistence that schools provide instruction in understanding language, symbolic form, and media.

Neil Postman[1980s] became a leading critic of television and the electronic media, appearing on television with increasing frequency to deliver his humanistic critique of the medium and its impact on human affairs. *Teaching as a Conserving Activity* took its place as the first of Postman's television trilogy, followed by *The Disappearance of Childhood* in 1982. There he noted that television reveals the secrets that we previously had kept from children as they sat, sequestered in the schoolroom; in revealing all, he argued that television blurs the distinction between childhood and adulthood characteristic of print culture. The third book, published in 1985, became his most powerful and widely acclaimed work. Entitled *Amusing Ourselves to Death: Public Discourse in the Age of Show Business*, he wrote again about crazy and stupid talk and the demeaning of meaning, this time as consequences of our wholesale adoption of television technology. Serious subjects, such as news, politics, religion, and education, become trivialized on television, he explained, because the medium's bias favors entertaining formats that emphasize images and immediacy. Postman earned the 1987 George Orwell Award for Clarity in Language from the National Council of Teachers of English on the strength of this analysis.

Postman remained an outspoken critic of the electronic media as he drew on semiotics to support a call for banning beer advertising from television in a controversial report prepared for the American Automobile Association Foundation for Traffic Safety, coauthored by Christine Nystrom, myself, and Charles Weingartner, and released in 1987; opposed the introduction of cameras into the courtroom as a member of a New York State Advisory Committee charged in 1988 with considering the innovation; and collaborated with newscaster Steve Powers to demystify broadcast journalism in the 1992 book, *How to Watch TV News*. Postman included additional critical essays on media (some of which first appeared in *ETC*) in his 1988 anthology, *Conscientious Objections*, along with essays on education and language, and a profile of Alfred Korzybski, which began with the preamble:

> In 1976, I was appointed editor of *ETC: The Journal of General Semantics*. For ten years, I served in that capacity, and with each passing year, my respect for Alfred Korzybski increased and my respect for those academics who kept themselves and their students ignorant of his work decreased. I here pay my respects to a unique explorer, and by implication mean to express my disdain for those language educators who steep their students in irrelevancies and who believe that William Safire and Edwin Newman have something important to say about language. (p. 136)

Neil Postman[1990s] expanded his critical focus to include information technologies such as the computer and the Internet. In 1992 he published *Technopoly: The Surrender of Culture to Technology*, in which he explained that we accept technology into our lives automatically and uncritically, allowing it to penetrate every sector of American society and govern every aspect of human activity. This prompted some to label Postman a neo-Luddite, although he mainly argued for giving more thought to the unforeseen and negative consequences of technology, and for maintaining values, ethics, and social institutions independent of the technological imperative. Moreover, even when he took as his theme the hardware of technology, he never lost sight of the paramount importance of the software of language, as *Technopoly* included chapters on "Invisible Technologies" and "The Great Symbol Drain." Similar critiques took the form of short pieces such as "Cyberspace, Shmyberspace" published in 1996.

Postman shifted his focus from media and technology to broader cultural issues in his final two major works. In 1995 he published *The End of Education*, in which he discussed the decline of our common culture and shared set of beliefs, a condition he had previously diagnosed as brought on by the electronic media and technopoly. Postman argued that under such conditions, public education could not maintain its vitality, nor even its viability. In 1999, as a response to President Clinton's rhetorical call to build a bridge to the 21st century, Postman gave us *Building a Bridge to the Eighteenth Century*. There he argued that we needed to bring with us into the new millennium the rationality of the Enlightenment (which he had earlier identified as a product of print culture). In both works, he remained steadfast in arguing that understanding language, media, and technology would go a long way towards solving our social ills. In *Building a Bridge to the Eighteenth Century*, Postman also criticized poststructuralist and postmodernist theorists such as Jacques Derrida and Roland Barthes for claiming, in effect, that since "the map is not the territory," the territory must not exist:

> If postmodernism is simply skepticism elevated to the highest degree, we may give it muted applause. The applause must be muted because even skepticism requires nuance and balance. To say that all reality is a social construction is interesting, indeed provocative, but requires, nonetheless, that distinctions be made between what is an unprovable opinion and a testable fact. And if one wants to say that "a testable fact" is, itself, a social construction, a mere linguistic illusion, one is moving dangerously close to a kind of Zeno's paradox. One can use a thousand words, in French or any other language, to show that a belief is a product of habits of language—and graduate students by the carload can join in the fun—but blood still circulates through the body and the AIDS virus still makes people sick and the moon is not made of green cheese. (p. 78)

Postman disliked postmodernism for its mistaken view of language and symbolic form, and also for the way that postmodernists and related cultural theorists used language in their writing, for their lack of clarity, dependence of jargon, convoluted sentence structure, etc. Reflecting on his career as an academic, he stated that

> if an academic has anything interesting or useful to say, I believe he or she has a responsibility to say it to fellow citizens. And, of course, to say it in a way that will capture and hold their interest. It is something of a minor tragedy that so many brilliant

academics I know—people who have a great deal to say of interest to their fellow citizens—have been conditioned to write in a way, as Shakespeare said it, that no human ear can endure to hear. (Gencarelli, Borisoff, Chesebro, Drucker, Hahn, & Postman, 2001, p. 134)

Postman taught his students that clarity is courage, and that we could achieve clarity by gaining an understand of language, symbolic form, and technology through the study of general semantics and media ecology.

Neil Postman[2000s] had served as Chair of New York University's Department of Culture and Communication for over a decade, had held the rank of University Professor since 1993, and the Paulette Goddard Chair of Media Ecology since 1998. He had continued to develop his ideas about media ecology in short pieces such as "The Humanism of Media Ecology," his keynote address delivered at the first Media Ecology Association convention. And he rejoined the Editorial Board of *ETC* in 2003.

In a 1994 article I published in *ETC* about Neil Postman, I characterized him as a "defender of the word" (Strate, 1994, p. 163),[*] and I believe that this best sums up his life's work. He labored to defend the word from the external threats posed by the proliferation of media images, and the technological drive to reduce all things to numbers. And he worked to defend the word against the internal threat posed by the abuse and misuse of language. Neil Postman[1931-2003] stands now with Korzybski, Hayakawa, McLuhan, et al, as writers, scholars, and educators for the ages.

---

[*] Reprinted in this anthology as Chapter Five.

Chapter Seven
Paradox Lost

L inda Elson was a friend, a colleague, and a brilliant scholar, and her untimely death was both a personal loss and a loss for the field of media ecology in its entirety. We are fortunate, however, that she left us with this study (Elson, 2010), which represents her life's work and, it should be readily apparent, a demonstration of the promise that was cut short by illness. But more importantly, this work also represents a valuable contribution to scholarship in general, especially in the field of communication and the study of symbols and logic, and to general semantics, general systems theory, relational communication, the study of humor and, of course, the field of media ecology.* Those not familiar with media ecology as an intellectual tradition might be forgiven if they fail to see a connection between this present work and the study of media as environments, since Elson does not concern herself with media as they are typically understood, e.g., television, radio, motion pictures, sound recordings, newspapers, magazines, the internet, etc. In other words, her focus is not on the media of mass communication, nor specifically on the media of interpersonal communication such as the postal service, telegraph, telephone, or the various forms of text messaging, e-mail, and online chat. Even broadening the concept of media, as Marshall McLuhan (2003) was known to do, to include all forms of technology from automobiles and clocks to clothing and architecture would not be sufficient, as Elson is not interested in media as gadgets, mechanisms or tools, nor with communication in a material sense. Rather, her focus is on form, and specifically symbolic form, a topic of great interest to media ecologists, and an area of study that constitutes one of the main foundations of our field.

Of Signs and Symbols

To understand the centrality of Elson's study to media ecology, it is worthwhile to review the origins of our intellectual tradition. As I have explained elsewhere (Strate, 2006), media ecology has no single point of origin, and no one founder, but rather coalesced largely during the 20th century as several different strands of scholarship intersected and began

---

* This chapter is taken from the Introduction I wrote for *Paradox Lost: A Cross-Contextual Definition of Levels of Abstraction*, by Linda Elson (2010), edited by Alan Ponikvar.

to inform one another. One of the strands, and many would argue the most important of them all, is the study of symbols, language, and human behavior. It would certainly be difficult to discount the significance of the nineteenth century philosopher, Charles Saunders Peirce (1991), who coined the term *semiotics*, defined as the science of signs, as part of his overall project in pragmatism. Peirce's semiotics breaks the overarching category of signs down into three different subcategories, the index, the icon, and the symbol, each one characterized by different signifiers associated with different signifieds (the signifier roughly corresponding to the medium, and the signified to the message). And in emphasizing the categorical differences among index, icon, and symbol, Peirce prepares the way for McLuhan's emphasis on the differences between different types of media, and the idea that each medium has its own bias or favors a particular type of content. It is therefore not surprising to find that contemporary media ecologists such as Paul Ryan (1993) and Aquiles Esté (1997) have explored various ways of synthesizing Peirce's semiotics with McLuhan's approach to understanding media. Similarly, Régis Debray, the Communist revolutionary turned media theorist, introduces the term *mediology* as an alternative to semiology, replacing signification with mediation, in *Media Manifestos* (1996), and *Transmitting Culture* (2000). In this way, Debray combines the perspective of McLuhan and other media ecology scholars with the French semiological tradition of Ferdinand de Saussure (i.e., *Course in General Linguistics*, 1983), which evolved along similar lines to Peirce's semiotics.

Of the different types of signs, it was the symbol that dominated the attention of 20th century scholars, as they focused on understanding the nature and functioning of the symbol as the basis of language, and more generally of human communication, consciousness, and culture. For example, in the discipline of English or literary studies, I.A. Richards collaborated with Charles Ogden, the inventor of Basic English (a version of the language stripped down to 850 words in order to facilitate international communication) on *The Meaning of Meaning* (Ogden & Richards, 1923), where they discussed the structure and function of language and symbolic communication. Symbols, they argued, may reflect some aspect of reality, but they may also refract what is real due to the inherent bias of the symbolic form or medium involved:

We have spoken . . . of reflection and refraction by the linguistic medium. These metaphors if carefully considered will not mislead. But language, though often spoken of as a medium of communication, is best regarded as an instrument; and all instruments are extensions, or refinements, of our sense organs. The telescope, the telephone, the microscope, the microphone, and the galvanometer are, like the monocle or the eye itself, capable of distorting, that is, of introducing new relevant members into the contexts of our signs. And as receptive instruments extend our organs, so do manipulative instruments extend the scope of our motor activities. When we cannot actually point to the bears we have dispatched we tell our friends about them or draw them; or if a slightly better instrument than language is at our command we produce a photograph. The same analogy holds for the emotive uses of language: words can be used as bludgeons or bodkins. But in photography it is not uncommon for effects due to the processes of manipulation to be mistaken by amateurs for features of the objects depicted. Some of these effects have been exploited by experts so as greatly to exercise the late Sir Arthur Conan Doyle and his friends. In a similar fashion language is full of elements with no representative or symbolic function, due solely to its manipulation; these are similarly misrepresented or exploited by metaphysicians and their friends so as greatly to exercise one another—and such of the laity as are prepared to listen to them. (p. 98)

What Ogden and Richards refer to as "effects due to the process of manipulation" would later come to be called the biases of a medium. And in referring to language as an instrument, Ogden and Richards are categorizing "the linguistic medium" as a technology. Their reference to media as extensions of the senses would later be echoed by Marshall McLuhan, who studied under Richards, and the New Criticism that Richards championed had a profound influence on Walter Ong and Neil Postman as well as McLuhan. In his collaboration with Ogden, and in his own works such as *Practical Criticism* (1929), and *The Philosophy of Rhetoric* (1936), Richards stresses the centrality of understanding language and symbolic communication to the study of literature, and to all of human knowledge.

Languages Old and New

In a somewhat different but related vein, Edward Sapir, an anthropologist specializing in linguistics, emphasized the categorical differences among the myriad specific languages spoken by human beings. Different languages, he argued, are associated with different worldviews, different modes of thought, understanding, and perception, different social constructions of reality. Sapir presented this perspective, sometimes referred to as linguistic relativism, along with a number of other ideas basic to the media ecology intellectual tradition in his popular 1921 book, *Language: An Introduction to the Study of Speech*:

> Languages are more to us than systems of thought transference. They are invisible garments that drape themselves about our spirit and give a predetermined form to all its symbolic expression. When the expression is of unusual significance, we call it literature. Art is so personal an expression that we do not like to feel that it is bound to predetermined form of any sort. The possibilities of individual expression are infinite, language in particular is the most fluid of mediums. Yet some limitation there must be to this freedom, some resistance of the medium. In great art there is the illusion of absolute freedom. The formal restraints imposed by the material—paint, black and white, marble, piano tones, or whatever it may be—are not perceived; it is as though there were a limitless margin of elbow-room between the artist's fullest utilization of form and the most that the material is innately capable of. The artist has intuitively surrendered to the inescapable tyranny of the material, made its brute nature fuse easily with his conception. The material "disappears" precisely because there is nothing in the artist's conception to indicate that any other material exists. For the time being, he, and we with him, move in the artistic medium as a fish moves in the water, oblivious of the existence of an alien atmosphere. No sooner, however, does the artist transgress the law of his medium than we realize with a start that

there is a medium to obey.

Language is the medium of literature as marble or bronze or clay are the materials of the sculptor. Since every language has its distinctive peculiarities, the innate formal limitations—and possibilities—of one literature are never quite the same as those of another. The literature fashioned out of the form and substance of a language has the color and the texture of its matrix. The literary artist may never be conscious of just how he is hindered or helped or otherwise guided by the matrix, but when it is a question of translating his work into another language, the nature of the original matrix manifests itself at once. All his effects have been calculated, or intuitively felt, with reference to the formal "genius" of his own language; they cannot be carried over without loss or modification. Croce is therefore perfectly right in saying that a work of literary art can never be translated. Nevertheless literature does get itself translated, sometimes with astonishing adequacy. This brings up the question whether in the art of literature there are not intertwined two distinct kinds or levels of art—a generalized, non-linguistic art, which can be transferred without loss into an alien linguistic medium, and a specifically linguistic art that is not transferable. I believe the distinction is entirely valid, though we never get the two levels pure in practice. Literature moves in language as a medium, but that medium comprises two layers, the latent content of language—our intuitive record of experience—and the particular conformation of a given language—the specific how of our record of experience. Literature that draws its sustenance mainly—never entirely—from the lower level, say a play of Shakespeare's, is translatable without too great a loss of character. If it moves in the upper rather than in the lower level—a fair example is a lyric of Swinburne's—it is as good as untranslatable. Both types of literary expression may be great or mediocre. (pp. 221-223)

It was Sapir's student, Benjamin Lee Whorf (1956), who fleshed out the concept of linguistic relativism, particularly through his studies of the languages of Native Americans such as the Hopi and Navajo. Dorothy Lee (1959) also championed this approach, in addition to exploring orality-literacy distinctions. Other anthropologists, such as Leslie White (1959), have stressed the leading role that all forms of symbolic communication play within a culture, while Edward T. Hall (1959) essentially defines culture as a system of communication itself. Hall extends what is sometimes known as the Sapir-Whorf Hypothesis, or the Sapir-Whorf-Lee hypothesis to cover culture in its entirety as a communication system, rather than just language, while his fellow anthropologist, Edmund Carpenter (1960), stretches the hypothesis to apply to all forms as media:

> English is a mass medium. All languages are mass media. The new mass media—film, radio, TV—are new languages, their grammars as yet unknown. Each codifies reality differently; each conceals a unique metaphysics. (p. 162)

Carpenter goes on to explain

> Each medium, if its bias is properly exploited, reveals and communicates a unique aspect of reality, of truth. Each offers a different perspective, a way of seeing an otherwise hidden dimension of reality. It's not a question of one reality being true, the others distortions. One allows us to see from here, another from there, a third from still another perspective; taken together they give us a more complete whole, a greater truth. (p. 162)

McLuhan (2003) takes things a step further by suggesting that all media, including all forms of technology, are translators, metaphors, and therefore, a kind of language. This understanding can be traced back to his doctoral research on the trivium, with a particular focus on grammar, in its original sense as the study of language and poetics—this work was recently published under the title of *The Classic Trivium: The Place of Thomas Nashe in the Learning of His Time* (McLuhan, 2006). Louis Forsdale (1981) was one of many scholars who have argued that McLuhan's understanding of media is essentially an extension of the Sapir-Whorf Hypothesis; it is also worth noting that Forsdale's students, Neil Postman and Charles Weingartner, were trained in linguistics, made it the subject of their first book, and incorporated it into their later work, a point I will return to subsequently. More recent efforts to build on Carpenter's (1960) call to understand the new languages of media include Robert Logan's in *The Fifth Language* (1997) and *The Sixth Language*

(2000), and Frank Zingrone's *The Media Symplex* (2001), both of which also incorporate systems perspectives (see below), not to mention Lev Manovich's film-based approach in *The Language of New Media* (2001).

Returning specifically to linguistics, the contemporary successors to Sapir and Whorf are George Lakoff and Mark Johnson, who specialize in the study of metaphor. Their perspective is that metaphor is much more than a literary or rhetorical device, but rather an intrinsic element of language that has a strong influence on the way we view and experience the world. Lakoff and Johnson established this perspective in *Metaphors We Live By* (1980), and followed up with a study of poetic metaphor, *More Than Cool Reason* (1989), and *Philosophy in the Flesh: The Embodied Mind and its Challenge to Western Thought* (1999). As the title implies, they see the human body as the basis for metaphor (and mind), as it is our primary point for making comparisons. Johnson also studies this in *The Body in the Mind* (1987), and Lakoff in *Women, Fire, and Dangerous Things* (1987), where he discusses how categories in thought are based on associations rather than abstractions. Taking this approach a step further, Raymond Gozzi, Jr. has combined the Lakoff and Johnson approach with McLuhan and Postman's media ecology in *The Power of Metaphor in the Age of Electronic Media* (1999). In addition to presenting numerous case studies of popular metaphor in contemporary American culture, Gozzi argues that metaphor has resurfaced as a dominant mode of communication in the electronic media environment, having been suppressed in literate cultures. This is a positive development in Gozzi's estimation, as he considers metaphor to represent the creative side of language. The same is true of neologism, and Gozzi explores this phenomenon as a basic form of linguistic creativity and as a reflection of the United States as a technological society in *New Words and a Changing American Culture* (1990).

The relationship between language and consciousness, and the extent to which thought is based on the medium of speech, has been explored by Russian psychologists Lev Vygotsky in *Thought and Language* (1986) for example, and by Alexander Luria in works such as *Language and Cognition* (1981), although their concern was with language in general as a medium of communication, much like Noam Chomsky (see, for example, *Language and Mind*, 1972). For the French anthropologist, Claude Lévi-Strauss, the structure of our symbolic forms mirrors the structure of our minds, a point he discusses at length in *Structural Anthropology* (1967). In particular, the bilateralism of brain and body is mirrored in the structure of language, which he believes to be based on binary oppositions (e.g., good and evil, life and death, nature and culture). As he explains in works such as *The Raw and the Cooked* (1969), this is also the structure of myth as a symbolic form, whose purpose it is to mediate contradictions. This requires a third term to perform the function of mediation of opposites. While mediation here is used in a somewhat different sense than it might be in reference to media of communication, Lévi-Strauss does deepen our understanding of mediation as the function carried out by a medium. Moreover, his notion that fire is the mediating agent that transforms nature (the raw) into culture (the cooked) suggests the basic unity of fire as *technology*, as *symbol* (mediating term), and as the defining characteristic of *culture*.

Philosophies of Symbolic Form

Of particular relevance to Elson's research are the philosophical studies of symbolism, language, and logic carried out by Alfred North Whitehead and Bertrand Russell, notably in their three volume *Principia Mathematica* (1925-1927). In particular, Elson's work is rooted in Whitehead and Russell's theory of logical types, which distinguishes between different levels of symbolic communication, a lower level on which you find the members of a group, and the higher level of the group itself. Groups may, however, be members of other groups that exist on a still higher level, groups of groups, and so on *ad infinitum*. In formalizing the idea that a category exists on a higher level of abstraction than the items it contains, and that a group cannot be a member of itself, Whitehead and Russell helped not only to resolve logical paradoxes, but also to set the stage for the parallel distinction between the medium and the message. Ludwig Wittgenstein extended Whitehead and Russell's line of inquiry with his *Tractatus Logico-Philosophicus* (1961), and his famous observation, "whereof one cannot speak, thereof one must be silent" (p. 189), is very much a statement about the ways in which a medium defines and delimits its messages. Wittgenstein's struggle with understanding the medium of language is reflected in this early work, and also in his *Philosophical Investigations* (1963) where he comes to view language as a game, thereby emphasizing medium over meaning.   Also of significance is Douglas Hofstadter's return to the theory of logical types in *Gödel, Escher, Bach* (1979), using the term recursion to refer to self-reference, a concept now commonplace in computer programming; Hofstadter goes so far as to suggest that recursion forms the basis of consciousness, as self-consciousness.

Russell, Whitehead, and Wittgenstein focus almost entirely on language as a symbol system, with special emphasis on logic and the use of propositions, statements that can be proven true or false based on the available evidence.   Coming out of the same philosophical tradition, but also drawing on the categorical distinctions of Peirce, and Ernst Cassirer's (1953) philosophy of symbolic forms, Susanne Langer represents a key transitional figure in the development of the media ecology intellectual tradition, as she broadened the definition of symbol to include art, music, ritual, and even sense perception in her key work, *Philosophy in a New Key* (1957):

> The abstractions made by the ear and the eye—the forms of direct perception—are our most primitive instruments of intelligence. They are genuine symbolic materials, media of understanding, by whose office we apprehend a world *of things*, and of events that are the histories of things. (p. 92; emphasis in the original).

This basic type of symbolism she terms presentational, as opposed to what she calls discursive symbols, a category that includes language and mathematics, and is associated with rationality, the ability to form statements that are propositional (that can be evaluated empirically), and broken down into discrete units (e.g., words, numbers).   Each type of symbol or medium has its own bias, so that while discursive symbols are well suited for producing logical statements, "language is a very poor medium for expressing our emotional nature" (p. 100).   Presentational symbols, on the other hand, cannot be used to form propositions that can be proven true or false, cannot be broken down into distinct units, but are well suited to understanding and conveying feeling and emotion. Moreover, she suggests that

> there may well be many special regions, to one or another of which the medium of one art is more suited than that of another

for its articulate expression. It may well be, for instance, that our physical orientation in the world—our intuitive awareness of mass and motion, restraint and autonomy, and all characteristic feeling that goes with it—is the preëminent subject matter of the dance, or of sculpture, rather than (say) of poetry; or that erotic emotions are more readily formulated in musical terms. I do not know, but the possibility makes me hesitate to say categorically, as many philosophers and critics have said, that the import of all the arts is the same, and only the medium depends on the peculiar psychological or sensory make-up of the artist, so that one man may fashion in clay what another renders in harmonies or in colors, etc. The medium in which we naturally conceive our ideas may restrict them not only to certain forms but to certain fields, howbeit they all lie within the verbally inaccessible field of vital experience and qualitative thought. (p. 258)

In this tentative and roundabout fashion, Langer is essentially saying that the medium is the message. She more fully and definitively explores the particular meaning of different forms or media in the sequel to *Philosophy in a New Key*, *Feeling and Form* (1953), and in her three volume *Mind: An Essay on Human Feeling* (1967, 1972, 1982).

General Semantics and General Systems

Certainly in line with Peirce's pragmatism is the practical approach known as general semantics, founded by Alfred Korzybski; his magnum opus, *Science and Sanity: An Introduction to Non-Aristotelian Systems and General Semantics*, was originally published in 1933, with a fifth edition in print as of 1993. General semantics was popularized by S. I. Hayakawa's *Language in Thought and Action*, originally published as *Language in Action* in 1939; a fifth edition co-authored by his son, Alan Hayakawa was published in 1990. Among the numerous other works explaining and elaborating on Korzybski's system are economist Stuart Chase's *The Tyranny of Words* (1938) and speech pathologist Wendell Johnson's *People in Quandaries* (1946). Following Whitehead and Russell, Korzybski is concerned with the relationship between language and the reality it is believed to represent, and argues that there are sources of error inherent in language itself, especially as expressed through Aristotelian logic. As an alternative, he put forth three non-Aristotelian principles of thought: the principle of non-identity, that words are not the things they represent, and that identity relationships only exist in symbol systems, not in nature; the principle of non-allness, that it is never possible to say all that there is to say about any particular phenomenon, that some things cannot be expressed in words or symbols, and that our capacity to create symbolic representations of reality will always leave out details; and the principle of self-reflexiveness, that while some words make reference to some aspect of reality, there also are words that refer to other words, and words that refer to those words, and so on (as there can be statements about statements, questions about questions, etc.). This last principle is derived from the theory of logical types, so that words that refer to words can be understood as a category, and the members of that group are words that refer to things in the world; words that refer to words would, in turn, be members of the category of words that refer to words that refer to words, and so on. Moreover, self-reflexiveness, along with the other two non-Aristotelian principles, are a by-product of the process of abstracting, as symbolic communication allows us to move from the more concrete to higher and higher levels of abstraction.

In a manner distinct from but paralleling Langer, general semanticists move away from the linear view of the world associated with Aristotelian logic, a view that other media ecology scholars such as Eric Havelock (1963, 1976, 1978, 1986), Walter Ong (1967, 1971, 1977, 1982, 2002), Jack Goody (1977, 1986, 1987, 2000), and McLuhan (1962, 2003; McLuhan & McLuhan, 1988) himself associate with the transition from orality to literacy. A similar shift from a linear, cause-and-effect orientation to one that is circular in orientation accompanied Norbert Wiener's (1950, 1961) introduction of cybernetics, the science of control, which emphasized the essential role that feedback plays in using information to maintain equilibrium. Wiener extends the notion of information theory developed by Claude Shannon (Shannon & Weaver, 1949), and Gregory Bateson further developed Shannon's work and Wiener's approach, famously defining information as "a difference which makes a difference" (p. 453), which expresses the same emphasis on categorical differences that can be found in Peirce's semiotics; this is a good way of describing media ecology's main concern as well, occupying a middle ground between the universalism of modernist theoretical formations and the particularism of many contemporary cultural theorists. Bateson brought together cybernetics and semantics, along with communication, anthropology, psychiatry, biology, and ecology; in his collected essays, entitled *Steps to an Ecology of Mind* (1972), Bateson proposes "a new way of thinking about *ideas* and about those aggregates of ideas which I call 'minds.' This way of thinking I call the 'ecology of mind,' or the ecology of ideas" (p. xv, emphasis in the original). And he continued to explore the common ground between psychology and biology in the recently reprinted *Mind and Nature: A Necessary Unity* (2002), while his coauthored book *Communication* (Ruesch & Bateson, 1951) remains influential in both the field of communication and the practice of psychotherapy.

Gregory Bateson was also a pioneer of systems theory, an approach that views all phenomena, not just language and symbols, as members of a group, or system. The distinction between the members of a category and the category itself is reflected in the systems view that the whole is greater than the sum of its parts. And the concept of self-reflexiveness is mirrored in the notion that a system can be a part of a larger system, and that system part of a larger system still, and so on. Thus, another pioneer of systems theory, Ludwig von Bertalanffy, named his version of the theory after general semantics with *General System Theory* (1969), and the variation, *general systems theory*, has also been used; the simpler form of *systems theory* or *systems view* is the most commonly used term, however as in the work of another pioneer in this area, Ervin Laszlo, e.g., *The Systems View of the World: The Natural Philosophy of the New Developments in the Sciences* (1972); a revised edition is published under the title *The Systems View of the World: A Holistic Vision for our Time* (1996). Additionally, Fritjof Capra has produced a synthesis of the concepts of nonlinearity, systems, complexity, networks, and ecology in works such as *The Tao of Physics* (1975), *The Turning Point* (1982), *The Web of Life* (1996), and *The Hidden Connections* (2002).

For Bateson, the systems view was particularly useful when applied to psychotherapy, and he was associated with the Mental Research Institute in Palo Alto, California, founded by Don Jackson in 1959, and whose early staff included Don Weakland, Richard Fisch, and Paul Watzlawick. Watzlawick, alone and in collaboration with others, built upon Bateson's foundation, introducing new approaches to therapy (e.g., brief therapy, family therapy), and establishing a number of key concepts in communication theory. The major work produced by Watzlawick is *Pragmatics of Human Communication* (1967), co-authored by Janet Bavelas (née Beavin), and Don Jackson. It is here that the first axiom of communication, "one cannot not communicate," is put forth. Here too, the important contrast is drawn between digital and analogical codes of communication, a binary opposition derived from computing that allows for a broad division between means or media of communication: digital would include most forms of language as well as all forms of number, and roughly corresponds to Langer's category of discursive symbols, while analogic encompasses most types of nonverbal communication, including pictures and music, and corresponds to Langer's notion of presentational symbolism.

Watzlawick *et al* also distinguish between two levels of communication, the content level, on which regular communication occurs, communication about the world in other words, and the relationship level, which involves communication about communication, or *metacommunication*; this amounts to another way to express the distinction between the parts of a group/system and the group/system itself, as well as the concept of self-reflexiveness. The distinction between content, which we generally pay attention to, and relationship, which we tend to ignore and therefore becomes an invisible environment, can be understood as another aspect of the distinction between content and medium. In other words, relationships are a type or aspect of media, and different media represent different types of relationships. For Watzlawick, the study of relationships is based on systems theory, and he presents his perspective on systems, which he associates with group theory in mathematics, in *Change* (1974), co-authored by John Weakland and Richard Fisch. Watzlawick also argues for the social construction of reality in *How Real is Real?* (1976), where he distinguishes between first order reality, which is about the physical properties of the world, and second order reality, which refers to the meaning and value that we assign to things. And he has returned to many of these same themes in subsequent works, *The Situation is Hopeless, But Not Serious* (1983), *Ultra-Solutions* (1988), and *Münchhausen's Pigtail* (1990). Watzlawick and his colleagues are sometimes known as the Palo Alto Group, which also include Bateson, Hall, Erving Goffman, and Ray Birdwhistel.

At this juncture, it would be worthwhile to note the continued development of systems theory, and the introduction of the systems concept of autopoiesis or self-organization, by the Chilean biologists Humberto Maturana and Francisco Varela in *Autopoiesis and Cognition* (1980), and explained for the general reader in *The Tree of Knowledge* (1992). Their view that self-organization involves closure against the system's environment supports radical positions on the social construction of reality. Moreover, Maturana and Varela's approach to systems theory provides the basis for the work of the German sociologist, Niklas Luhmann. In studies such as *The Differentiation of Society* (1982), *Ecological Communication* (1989), *Essays on Self-Reference* (1990), *Social Systems* (1995), *Art as a Social System* (2000a), and *The Reality of the Mass Media* (2000b), Luhmann puts forth a view of society as a system whose parts are not individuals or institutions, but acts of communication. Drawing on Ong (1967, 1982), Havelock (1963, 1976), and Elizabeth Eisenstein (1979), Luhmann sees media history as a process that generates increasing amounts of information, which in turn leads to the development of an increasingly more complex society. Complexity, in this instance, refers to the process by which systems generate their own internal subsystems (e.g., legal, political, economic, and educational subsystems), each of which becomes relatively autonomous within the context of the larger system. Influenced by general semantics and related approached to symbolic communication, Luhmann emphasizes the role of binary coding in maintaining the boundaries of the subsystems, and the social system as a whole, and sees the mass media as constructing a simplified and self-referential conception of the environment. Additionally, Jeremy Campbell has produced a popular summary of cybernetics, information theory, and the systems approach that incorporates media ecology's emphasis on communication and language in his book, *Grammatical Man* (1982).

Postman and Media Ecology

The scholarship of Neil Postman and his colleagues, culminating in the formalization of media ecology as a field of inquiry, brought together all of these intertwining strands, together with the study of media and technology. Postman wrote about the relevance of understanding language, symbols, and media for English education as early as 1961 in a book commissioned by the National Council of Teachers of English, entitled *Television and the Teaching of English*. He also advocated language education, emphasizing the communication process or medium over content, as an alternative to traditional approaches to grade school English, with its focus on prescribing proper grammar, spelling, and other facets of elite culture. Collaborating with Charles Weingartner, Postman elaborated on this argument in *Linguistics: A Revolution in Teaching* (Postman & Weingartner, 1966). With the understanding that media constitute our new languages (Carpenter & McLuhan, 1960), Postman and Weingartner integrated the two positions to produce their highly successful *Teaching as a Subversive Activity* (1969), which was particularly popular within the educational reform movement of the 1960s. Reflecting McLuhan's criticism of print-based schools as outmoded and obsolescent, Postman and Weingartner called for new modes of education better suited to the age of electronic media, and in particular, called for a curriculum based on the "Sapir-Whorf-Korzybski-Ames-Einstein-Heisenberg-Wittgenstein-McLuhan-Et Al. Hypothesis . . . that language is not merely a vehicle of expression, it is also the driver; and that what we perceive, and therefore can learn, is a function of our languaging processes" (p. 101). *Teaching as a Subversive Activity* had a dramatic impact on the educational reform movement of that time, and remains influential to this day.

Further influenced by systems theory and Watzlawick's relational approach, Postman emphasized linguistics, semantics, and the study of interpersonal communication in his 1977 book, *Crazy Talk, Stupid Talk*; that same year, he began a 10 year term as editor of *ETC: A Review of General Semantics*. Postman moved into media criticism proper in 1979, the year he published *Teaching as a Conserving Activity*. Reversing himself from the position he had taken with Weingartner in *Teaching as a Subversive Activity* (1969), Postman argued for a thermostatic view, the term he used to refer to a cybernetic, feedback-based approach, in which schools would serve to counterbalance the prevailing biases of the culture, which by then were clearly with the electronic media. Arguing that schools and television are competing forms of education, Postman focuses on the key opposition between the word (both oral and literate, but reaching its highest form in print culture) and the image (which television makes predominant). This line of inquiry is then continued in *The Disappearance of Childhood* (1982), in which Postman argues that the concept of an extended childhood is a construction of print culture that has been destroyed by the leveling effect of the televised image, and in *Amusing Ourselves to Death* (1985), where he argues that our image culture trivializes serious discourse, e.g., news, politics, religion, and education. The emphasis on public discourse as a function of language, symbols, and media, continued to dominate his later works, including *Conscientious Objections* (1988), *Technopoly* (1992), *The End of Education* (1995), and *Building a Bridge to the Eighteenth Century* (1999).

Postman formally introduced the term "media ecology" in 1968, in an address delivered at the annual meeting of the National Council of Teachers of English (published under the title of "The Reformed English Curriculum" in 1970), and defined it as "the study of media as environments," and explains that its main concern is "how media of communication affect human perception, understanding, feeling, and value; and how our interaction with media facilitates or impedes our chances of survival. The word ecology implies the study of environments: their structure, content, and impact on people" (p. 161). And in *The Soft Revolution* (Postman & Weingartner, 1971), media ecology is described as "the study of transactions among people, their messages, and their message systems" (p. 139). Postman and his colleagues founded their doctoral program in media ecology at New York University in 1970, and Christine Nystrom wrote the first major dissertation on the subject, *Towards a Science of Media Ecology* (1973), where she draws a parallel between the development of media ecology on the one hand, and cybernetics and systems theory on the other, as both represent holistic, ecological approaches. Nystrom went on to join the media ecology faculty, and along with Postman, guided Linda Elson's doctoral research, and therefore this study.

Situated within this larger context of the media ecology intellectual tradition, this book takes its place alongside the works of Whitehead, Russell, Korzybski, Wittgenstein, Langer, Bateson, Watzlawick, and Postman. Elson's scholarship serves to improve our understanding of the medium of language and symbolic communication, and also very significantly, contributes to the formation of an eco-logic that ultimately has to serve as a philosophical foundation for our field and its scholarship. We can only be grateful that she was able to complete this study and pass it on to us, so that her unique ideas, insights, and discoveries can live on and continue to inform and inspire future generations.

Chapter Eight

The Ten Commandments and the Semantic Environment

The Ten Commandments and the Semantic Environment

n all of human history, few documents have been so widely disseminated and so deeply venerated as the Ten Commandments. In Judaism, they represent God's covenant with Israel, and the first of a total of six hundred and thirteen laws contained within the Five Books of Moses. Most of those laws were specific to the Jewish people, and therefore rejected by the early Christians who were interested in establishing a universal religion. But the Ten Commandments were in fact adopted by the Church, and have remained prominent through all of Christianity's schisms and divisions. Also, the followers of Mohammed included the Ten Commandments within Islam, incorporating a version into the Koran. Apart from religious traditions, the Ten Commandments also represent a philosophical system, specifically a system of ethics and, as one of the earliest examples of codified law, a symbol of legal systems and the ideal of justice. And while I think it is safe to say that the Ten Commandments are fairly well known throughout the world, exactly what people know about them remains in question. To quote one summary of a survey reported by UPI, "only 68 of 200 Anglican priests polled could name all Ten Commandments, but half said they believed in space aliens" (Ontario Consultants on Religious Tolerance, n.d.).

I myself had a bit of a close encounter with the way that some people think about the Ten Commandments recently, one that I would like to share with you. This came about in response to a post I made on a poetry blog I started on MySpace. I posted a "poem" there, although it did not have to be designated as such, under the title of "The Ten Commandments," followed by "As Told By God to Moses (A New Translation and Interpretation by Lance Strate)" (Strate, 2007). My intent was to come up with a new way to present the Ten Commandments that would boil them down to their basic values, and present them in a way that was short, lighthearted, and in tune with modern sensibilities. What I came up with was the following:

The Buck Stops Here!
Use Your Words,
But Choose Them With Care!
Give Us Both A Break,
And Your Parents, Too!
Respect Life,
Family, And
Labor As Well!
Be Honest, And
Be Content!

I received some nice feedback on this piece, some of it polite, some very positive, but what was particularly interesting to me was the fact that some people were not so happy with what I had done. They explained that they felt that the Ten Commandments *is* the Ten Commandments, and that the sacred text should not be tampered with. General semanticists will recognize this as an expression of Aristotelian thinking, based on the Law of Identity, and of course, a product of the tyranny of that pesky little word, *is*. For my part, I saw it as a marvelous teaching opportunity, and set about explaining why the Ten Commandments *is not* the Ten Commandments, that is, why the Ten Commandments does not constitute *a thing*, or more specifically, why there is no single definitive version of this text, but rather multiple variations.

One reason for variation is abbreviation. Popular representations of the Ten Commandments typically represent each commandment as a single sentence. And some of them are just one sentence, such as "You shall not steal," and "You shall not commit adultery." But some of the commandments are full paragraphs. For example, the commandment that is often represented as, "Remember the Sabbath day, to keep it holy," goes on to decree:

> Six days you shall labor, and do all your work, but the seventh day is a Sabbath to the Lord your God; in it you shall not do any work, neither you, nor your son, nor your daughter, your manservant, or your maidservant, nor your cattle, nor the sojourner who is within your gates; for in six days the Lord made heaven and earth, the

sea, and all that is in them, and rested the seventh day; therefore the Lord blessed the Sabbath day and hallowed it. (Exodus 20:7-10; Deuteronomy 5:11-14)

Beyond abbreviation, a greater source of variation is translation. I am fairly certain that none of the individuals who told me that the Ten Commandments *is* the Ten Commandments could speak or read Hebrew. What they thought of as the Ten Commandments is actually the English translation, and what all too many people fail to recognize is that every translation is, to a significant degree, an interpretation. For example, one of the most frequently cited variations in translation is the commandment popularly known as "You shall not kill," which is more precisely translated as, "You shall not murder." That is a difference that makes quite a bit of difference, as "murder" more clearly allows for self-defense, capital punishment, and the slaughter of animals for food. I should note that while translation is always problematic, it is especially so for a text written several thousand years ago. And further adding to the ambiguity of meaning is the fact that the Hebrew aleph-bet that the text is written in uses no vowels, and almost no punctuation marks. For reasons such as these, Judaism developed the tradition of Talmudic interpretation.

Along with abbreviation, translation, and interpretation, there are variations due to transmission. Simply put, the Ten Commandments were not written in stone. Well, actually, they were, at least according to the Bible, but the first set was destroyed by Moses after seeing the Israelites worshipping the golden calf, and the second set disappeared centuries later, as you no doubt know from viewing *Raiders of the Lost Ark*. Presumably, the text was copied onto papyrus scrolls at some point, and copies made of the copies, and so on self-reflexively. And it is well known that scribal copying was prone to corruption due to error or deliberate alteration, so it is highly likely that multiple versions were in circulation at one time, before the Bible was canonized. In fact, there are two different versions in the Bible, the first in Exodus (20:2-13), the second, with minor differences, in Deuteronomy (5:6-17). Even if you rule out the possibility of other alternative versions having existed at one time, the question remains as to which of the two authorized versions should be considered *the* Ten Commandments, the original in Exodus or the revised version in Deuteronomy?

Even more significant than all the others is the problem of enumeration. Simply put, while it is stated that there are indeed ten commandments, they are not numbered or punctuated in a way that clearly marks the beginning and end of each commandment. Consequently, different religious traditions have divided up the Ten Commandments in different ways, and there is no way to determine with absolute certainty how to punctuate the text. Perhaps the most significant difference exists between the Roman Catholic version, where their First Commandment includes the prohibition against graven images so that it is not a law unto itself, and most other traditions which identify that prohibition as the Second Commandment; most other traditions in turn consider the final set of sentences pertaining to coveting your neighbor's wife and neighbor's property as one commandment, while the Catholic tradition separates the two.

Allowing then for variations on account of abbreviation, translation, interpretation, transmission, punctuation, and enumeration, I want to proceed, despite the ambiguities, with the relatively common version of the Decalogue. Moreover, in taking up the topic of the Ten Commandments, I do not want to engage in debates as to the origins of the document. Regarding the question of God's authorship, that is a proposition that cannot be proven true or false based on the available evidence, and is therefore best set aside. As for other debates concerning the document's authorship and history, for our purposes I think it is sufficient to regard it as a product of antiquity, and not concern ourselves with questions such as whether it originated over three millennia ago, or only two and a half millennia past.

The Significance of Monotheism

With that preamble now complete, I want to begin by acknowledging that Neil Postman (1985) first suggested that we view the Ten Commandments as an effort to influence the semantic environment and the media environment of the ancient Israelites, especially in regard to the Second Commandment. But let me begin at the beginning, with the First Commandment, which says something along the lines of, "I am the Lord your God, who brought you out of the land of Egypt, out of the house of slavery. You shall have no other gods before Me" (Exodus 20:2;

Deuteronomy 5:6). The First Commandment is often thought to establish monotheism in a formal sense, or at the very least to be an important step in that direction. In contemporary terms, I think it comes across as unduly authoritarian, and it certainly stands in opposition to earlier religious and spiritual approaches known variously as polytheism, paganism, animism, shamanism, and the like, including the Goddess worship that Leonard Shlain (1998) champions. And so, we might well ask, what difference does it make to replace a multitude of deities with the One God of the Bible?

I think it is reasonable to think of the First Commandment as the foundation on which all of the others laws and commandments rest, the first principle as it were. And of course, it makes sense to begin by invoking a higher authority in order to invest the subsequent laws with authority and legitimacy, but that does not answer the question of why One God, why no God but God, why monotheism? I believe that the answer is that, if there is only One God, then there is also order: not competing forces; not gods in conflict with one another, each one vying for our loyalty, and the world serving as their battleground; and not the chaos of multiplicity. The Biblical story of Creation is the story of God making order out of chaos, and it is consistent with contemporary physics, information theory, cybernetics, and systems theory, and recent theories of chaos and complexity. God orders the world by way of differentiation, separating light from darkness, night from day, the sky from the earth from the waters, etc. The presence of One God is a guarantee that there is order out there, even when all we can perceive is randomness.

When we have One Creator, we therefore have One Creation, a world that is coherent and comprehensible. Albert Einstein said that, "God does not play dice with the universe." Well, why not? Simple! Because He has no one to play with! The gods of polytheism play games with the world, and with peoples' lives. The One God does not, and leaves nothing to chance. This also means that the world is open to inspection, to investigation, and ultimately to science. The world is an open book, orderly, and Lawful. Robert Logan (2004) has pointed to the establishment of codified law in ancient Israel as a necessary precondition for the study of natural law in the form of scientific inquiry. In addition to studying the universe, codified law also opens the door to the concept of universal human rights, indeed of rights for all living things. One God,

One Creation, One Humanity! And if there is one God, then He has ultimate responsibility, He says, "the buck stops here!" And you can argue with Him, and be angry at Him, but you can't go and get a second opinion, or appeal your case somewhere else; you can't find another god who will give you what you want. As an ethical principle, this means that we ourselves also have to be responsible, responsible to the One God, and also responsible to others, and to ourselves.

Abstraction vs. Idolatry

What I am suggesting is that in the First Commandment alone, there is an effort to establish a new way of thinking about the world, one that emphasizes logical coherence and systematic thinking, and above all, abstract thought. A number of media ecologists, including Harold Innis (1951, 1972), Neil Postman (1985), Robert Logan (2004), and Leonard Shlain (1998), have pointed to the relationship between the invention of the Semitic aleph-bet circa 1500 BCE, and the establishment of the first monotheistic religion. Simply put, members of oral cultures do not typically engage in the kind of abstract thinking that is characteristic of literate societies. And belief in one god is more abstract than belief in many, especially when the One God is conceived of as universal, omnipresent, almighty, all knowing, and invisible. The God of monotheism becomes the ultimate abstraction. And while general semanticists have been concerned over the past century with the problem of too much abstract thinking, several millennia ago the problem was that there was too little abstract thinking. And the solution was to encourage literacy. This is where Postman's (1985) insight about the Second Commandment comes in, as he argued that it discourages the concrete thinking associated with visual images, and encourages us to use our words:

> In studying the Bible as a young man, I found intimations of the idea that forms of media favor particular kinds of content and therefore are capable of taking command of a culture. I refer specifically to the Decalogue, the Second Commandment of which prohibits the Israelites from making concrete images of anything. "Thou shalt not make unto thee any graven image, any likeness of any thing that is in heaven above, or that is in the earth

beneath, or that is in the water beneath the earth." I wondered then, as so many others have, as to why the God of these people would have included instructions on how they were to symbolize, or not symbolize, their experience. It is a strange injunction to include as part of an ethical system *unless its author assumed a connection between forms of human communication and the quality of a culture.* We may hazard a guess that a people who are being asked to embrace an abstract, universal deity would be rendered unfit to do so by the habit of drawing pictures or making statues or depicting their ideas in any concrete, iconographic forms. The God of the Jews was to exist in the Word and through the Word, an unprecedented conception requiring the highest order of abstract thinking. Iconography thus became blasphemy so that a new kind of God could enter a culture. (pp. 8-9)

At this point, I should note that the Second Commandment says something like, "You shall not make for yourself a graven image, or any likeness which is in the heavens above, which is on the earth below, or which is in the water beneath the earth. You shall not prostrate yourself before them, nor worship them, for I, the Lord your God, am a jealous God, visiting the iniquity of the fathers upon the sons, upon the third and the fourth generations of those who hate Me. And I perform loving kindness to thousands of generations of those who love Me and to those who keep My commandments" (Exodus 20:3-5; Deuteronomy 5:7-9). The Second Commandment includes the prohibition against idol worship, a form of ritual associated with earlier approaches to religion and spirituality, and there is a polemic against idol worship that runs throughout the Hebrew Bible. For example, media ecologists such as Edmund Carpenter (Carpenter & Heyman, 1970) and Marshall McLuhan (2003) have quoted from the 115th Psalm of David (2-8), which seems to suggest something along the lines of McLuhan's famous adage, *the medium is the message:*

Wherefore should the nations say:

'Where is now their God?'

But our God is in the heavens;

On the Binding Biases of Time

Whatsoever pleased Him He hath done.

Their idols are silver and gold,

The work of men's hands.

They have mouths, but they speak not;

Eyes they have, but they hear not;

Noses have they, but they smell not;

They have hands, but they handle not;

Feet have they, but they walk not;

Neither speak they through their throat.

They that make them shall be like unto them;

Yea, every one that trusteth in them.

This can be understood as a critique of both visual imagery, and of technology, "the work of men's hands" as it were. But most importantly, it is a critique of the concrete. Idols are concrete, tangible. So are trees, mountains, bodies of water, the sky, the sun, moon, planets and stars, etc. Nature is concrete, and so is technology. As objects of worship, all of these are rejected by monotheism in favor of a transcendent concept of the divine that goes hand in hand with the universality of Law, and Justice, and human rights. We can therefore think of the Ten Commandments as representing an effort to raise people's consciousness to a new level. It may be that at one time people thought mostly in images, and when they thought in words it was so unusual that they imagined they were hearing voices from the outside, from supernatural beings, as Julian Jaynes (1976) has argued (see also Ong, 1982, on Jaynes). Or maybe people thought in words, but words tied to concrete images and experiences, not abstract concepts. Whatever the case may be, the problem with images or presentational symbols, to use Susanne K. Langer's (1957) term, is that we are more likely to confuse them with the "things" they are thought to represent, because they in some way resemble "things," as opposed to words, which are arbitrary and conventional. We are therefore more likely to respond to images with signal reactions, immediate, reflex reactions, rather than with delayed and reflective symbol reactions. We are more likely to engage in dead level abstracting, as Wendell Johnson (1946) put it, to remain on the same, low level of abstraction, rather than moving up and down the abstraction ladder, which is what general semanticists like Johnson recommend. And we are unable to construct logically coherent statements with images, let alone test their validity against evidence gleaned from sense perception. For all these reasons and more, it is possible to understand the First and Second Commandments as constituting an attempt to facilitate the evolution of consciousness within an entire culture.

Having laid the groundwork for a more abstract and systematic approach based on words over images, the Third Commandment presents us with the following admonishment: "You shall not take the name of the Lord, your God, in vain, for the Lord will not hold blameless anyone who takes His name in vain" (Exodus 20:6; Deuteronomy 5:10). Again, this comes across as almost dictatorial to our modern sensibilities, but it does get across a message that runs through many of the commandments: Be respectful! This is not a bad message, after all, but I also want to suggest to you that inherent in the Third Commandment is a more general rule, to the effect of, think before you speak. Once uttered, spoken words cannot be retrieved, and the damage that they might do cannot be entirely undone. Of course, there was a time when an oath taken in God's name was treated with the utmost seriousness by all, and given that attitude, the Third Commandment tells people to think things through before making promises you may not be willing or able to keep. In general, then, it is an effort to get people to delay their responses, be reflective, employ symbol reactions, and respect the power of language by choosing your words with care.

A New Sense of Time

I have already mentioned the Fourth Commandment, which goes something like, "Keep the Sabbath day to sanctify it, as the Lord your God commanded you. Six days may you work, and perform all your labor, but the seventh day is a Sabbath to the Lord your God; you shall perform no labor, neither you, your son, your daughter, your manservant, your maidservant, your ox, your donkey, any of your livestock, nor the stranger who is within your cities, in order that your manservant and your maidservant may rest like you. And you shall remember that you were a slave in the land of Egypt, and that the Lord your God took you out from there with a strong hand and with an outstretched arm; therefore, the Lord, your God, commanded you to observe the Sabbath day" (Exodus 20:7-9; Deuteronomy 5:11-13). And clearly, one of the goals of the commandment is to structure the temporal environment, and in fact to establish the week as a unit of time, working in tandem with the story in the Book of Genesis of how God created the world in six days and rested on the seventh. The week is the only calendar division that requires this sort of legitimization, I might add. By way of contrast, the day is based on the earth's rotation, the month on the cycles of the moon, and the seasons and years on the earth's revolution around the sun. In other words, marking the day, month, and year remains natural, and linked to the signal or index as a form of signification, but marking the week is artificial and places us firmly and unequivocally in the realm of the symbolic, and the abstract. Within this context, the Fourth Commandment also establishes order by differentiating between the profane time of the weekdays, and the sacred time of the Sabbath.

Of course, there is certainly much to be said for the idea of taking a break from labor, and taking some quiet time to engage in thought, meditation, and prayer. Donna Flayhan has made the point (on a number of occasions in personal communications and conference presentations) that, in the face of contemporary technological society and the constant bombardment of images and information that we experience, the idea of observing a Sabbath seems like a very good idea indeed, and this includes a Sabbath from leisure pursuits and entertainment as well as labor. I would also note that a longstanding assignment in introductory mass media classes has been asking the students to take a media fast, to abstain from using all forms of media, for a short period of time such as a day or week, and then writing about the experience. But to return to my basic point, the Fourth Commandment provides further reinforcement for both abstract thinking, and delayed, reflective, symbol reactions.

Respect and Ideal Communication

The Fifth Commandment reads something like, "Honor your father and your mother as the Lord your God commanded you, in order that your days be lengthened, and that it may go well with you on the land that the Lord, your God, is giving you" (Exodus 20:11; Deuteronomy 5:15). The message here is again one of respect, respect for others, for family, and for elders, which also suggests the importance of maintaining tradition, memory, and cultural continuity. From a general semantics point of view, this commandment reinforces the importance of time-binding.

The Sixth Commandment simply tells us, "You shall not murder" (Exodus 20:12; Deuteronomy 5:16) underlining the importance of having respect for life, which within this system is considered God's creation. The Seventh Commandment says, "And you shall not commit adultery" (Exodus 20:12; Deuteronomy 5:16). In the wake of the sexual revolution, this seems a bit incongruous following the prohibition against murder, and we have long since rejected the sort of situation associated with the scarlet letter. Be that as it may, this commandment can be understood as again asking us to have respect for others, and for the family. I think it interesting to note that marriage, an institution whose definition has become so contested nowadays, is present in the Ten Commandments here only in regard to its contravention. The Eighth Commandment reads, "And you shall not steal" (Exodus 20:12; Deuteronomy 5:16) . Once more, what is implied is that we need to have respect, in this instance respect for others' property, and more importantly I believe, respect for their labor. And taken together, the Sixth, Seventh, and Eight Commandments also tell us, once more, to be thoughtful, not to act on impulse, be it rooted in aggression, sexual drives, or some other form of desire. Again, we are told to delay our responses to people and things that we encounter in our environment, be reflective, and employ symbol reactions.

The Ninth Commandment brings us back to direct engagement with communication, as it states, "And you shall not bear false witness against your neighbor" (Exodus 20:12; Deuteronomy 5:16). In other words, it says that "honesty is the best policy," and requires that communication be accurate as well, in order to avoid inadvertent falsehood. In the parlance of general semantics, it tells us to make sure that our maps represent the territory as closely as humanly possible. I think it important to also note that this is exactly the ideal type of communication that serves as the basis of modern science, and also what Jurgen Habermas (1984-1987) refers to as the ideal speech act required for a truly democratic society.

On the Binding Biases of Time

Fundamentally, this Commandment requires us to operate with an extensional rather than intensional orientation, to again reference general semantics, that is, to set aside our emotions and desires, and to describe the world not as we want it to be, not colored by our preconceptions, but as much as possible to provide a factual, objective report free of judgments and inferences.

The Tenth Commandment concludes with something like, "And you shall not covet your neighbor's wife, nor shall you desire your neighbor's house, his field, his manservant, his maidservant, his ox, his donkey, or anything that belongs to your neighbor" (Exodus 20:13; Deuteronomy 5:17). This reinforces the commandments prohibiting adultery and theft, again requires the individual to be respectful of others, and asks people to be content with their lot. And once more, it asks us to think before we act, to take control over our thought processes, emotions and urges, to be conscious and aware of what we are doing. That is, this commandment also requires people to employ symbol reactions, to be reflective and delay their responses. It asks us to know ourselves, know our own minds, and take charge over ourselves. And more than any of the other commandments, it asks us to look inward, to examine and be conscious of our own motives and motivations, our own thoughts and emotions, our own process of abstraction.

By way of conclusion, let me note that my point in all this is not to suggest that media ecologists and general semanticists ought to get religion, or that the Ten Commandments ought to be the law of the land. I simply want to indicate the ways in which the Decalogue served not only to establish a new form of religion in the ancient world, but as a means of influencing the semantic environment and the media environment, and in doing so, as a means of influencing the evolution of culture and consciousness. As for the question of whether Moses was the first media ecologist, I leave that up to you.

Chapter Nine

Tolkiens of My Affection

T he title that I have taken for this chapter is something of a pun, as the name *Tolkien* is frequently mispronounced as Tolk*ehn*, and when written out looks like the word *token*. No doubt Professor J.R.R. Tolkien would have been able to explain the linguistic origins of this mispronunciation. He was, after all, a renowned philologist who held first the Rawlinson and Bosworth Chair of Anglo-Saxon, and later the Merton Chair of English Language at Oxford University. And perhaps, as an expert in linguistics, he would not have been too insulted to have his name conflated with the term *token*, given that *token* can be defined as a symbol or sign. And my intention is to present this essay as a symbol or sign of my affection for the author and his works. I suspect that *affection* is a term that is not used very frequently in serious scholarship, as we academics tend to traffic in thoughts, rather than emotions, in arguments and propositions rather than feelings and intuitions. But I take as my authority Susanne K. Langer, who in works such as *Philosophy in a New Key* (1957), *Feeling and Form* (1953), and *Mind: An Essay on Human Feeling* (1967, 1972, 1982) has championed the study of emotion in cognition and symbolic form. And when it comes to books like *The Lord of the Rings* (Tolkien, 1965a, 1965c, 1965d), there is no denying the powerful feelings that the novel evokes in so many readers.

Tolkien lovers exhibit the fervor of the spiritual convert, not the objectively distanced appreciation of the critical reader. This too can be disturbing to the serious scholar, unless *The Lord of the Rings* is framed as a religious narrative. And Tolkien did confess that the novel is "a fundamentally religious and Catholic work; unconsciously so at first, but consciously in the revision" (quoted in Shippey, 2000, p. 175). I should also mention that he wrote this in a letter to a Jesuit priest, but what is particularly interesting is what he went on to write:

> That is why I have not put in, or have cut out, practically all references to anything like 'religion', to cults or practices, in the imaginary world. For the religious element is absorbed into the story and the symbolism. (p. 175).

The result is something quite different than the religious fiction of Tolkien's fellow Inkling, C.S. Lewis. *The Lord of the Rings* does not point to any specific institution, belief-system, or practice. Rather, it presents us with a narrative that represents religious experience as a symbolic form, a narrative that evokes the general conception of religious experience in all of its varieties, as William James would have it. Along with *The Hobbit* (Tolkien, 1965b), we have a set of stories that take us from the familiar and profane world of the Shire, to alien landscapes and sacred spaces inhabited by wizards, elves, dwarves, nature spirits, dragons, wraiths, and demons. We have a hero's journey, but in place of Joseph Campbell's (1968) monomyth, we have Tolkien's multimyth, one for each member of the fellowship of the ring.

Frodo's quest is necessitated by Bilbo's earlier travels "there and back again," but in place of an adventure we have the solemn enactment of the scapegoat myth, as discussed by Kenneth Burke (1950). Sam's journey begins in service to his master, but ends with his mastering of himself. Merry and Pippin both go through a rite of passage from playful youth to mature leadership. Legolas and Gimli begin by championing their own races, but go on to transcend the limitations of species loyalty to become defenders of all life. Gandalf falls and rises, moving from life to death and back to life again, while Boromir's is the failed hero's journey, a failure of virtue followed by death and the final return and cremation of his body. And Aragorn's is the most traditional hero's journey, as he separates himself from his mundane existence as a ranger, faces many trials as his initiation, and returns as the King triumphant.

These stories represent spiritual journeys, but the journeys also represent our progress through the stages of life. We are all on a one-way trip to Mount Doom. And, we all hope for a resting place beyond the sea. We all must live our lives knowing that in the end we will meet death, and in that way as well as in many others we will fail. And we must find the courage to live with this knowledge, and to have faith and do the right thing even if we do not know the way, and the situation seems hopeless. The denial of death, and the discovery of the hero within all of us, is essential to the human psyche, as Ernest Becker (1971, 1973) explains.

It seems to me that these themes speak to us all the more powerfully following 9/11. When Peter Jackson's film of *The Fellowship of the Ring* was released just three months after the terrorist attacks, I could not help but be deeply moved to hear the exchange between Frodo and Gandalf taken from the book's second chapter, "The Shadow of the Past":

'I wish it need not have happened in my time,' said Frodo.

'So do I,' said Gandalf, 'and so do all who live to see such times. But that is not for them to decide. All we have to decide is what to do with the time that is given to us." (Tolkien, 1965a, p. 82)

Jackson rightly highlights these lines which, in the book are immediately followed by discussion of "the Enemy," significant no doubt but obscuring the universal meaning of Gandalf's first few sentences. There is the suggestion of a higher power, in the implication that someone other than ourselves decides about the times we are to live in. And there is the affirmation of free will within the limitations of a divinely ordered universe.

I have been discussing the religious quality of *The Lord of the Rings* to help to explain the strong emotion that many of us feel towards the book, and therefore its popularity. And if I seem too extreme in this, consider the recent study by Thomas Shippey, who held the Walter J. Ong Chair in the English Department of Saint Louis University. Published at the end of the twentieth century, the book is entitled *J.R.R. Tolkien*, and subtitled, *Author of the Century*. Now, if this seems mere hyperbole, here is how Shippey (2000) justifies the claim:

Late in 1996 Waterstone's, the British bookshop chain, and BBC Channel Four's programme *Book Choice* decided between them to commission a readers' poll to determine 'the five books you consider the greatest of the century'. Some 26,000 readers replied, of whom rather more than 5,000 cast their first place vote for J.R.R. Tolkien's *The Lord of the Rings*. Gordon Kerr, the marketing manager for Waterstone's, said that *The Lord of the Rings* came consistently top in almost every branch in Britain (105 of them), and in every region except Wales, where James Joyce's *Ulysses* took first place. The result was greeted with horror among professional critics and journalists, and the *Daily Telegraph* decided accordingly to repeat the exercise among its readers, a rather different group. Their poll produced the same result. The Folio Society then confirmed that during 1996 it had canvassed its entire membership to find out which ten books the members would most like to see in Folio Society editions, and had got 10,000 votes for *The Lord of the Rings*, which came first once again. 50,000 readers are said to have taken part in a July 1997 poll for the television programme *Bookworm*, but the result was yet again the same. In 1999 the *Daily Telegraph* reported that a Mori poll commissioned by the chocolate firm Nestlé had actually managed to get a different result, in which *The Lord of the Rings* (at last) only came second! But the top spot went to the Bible, a special case, and also ineligible for the twentieth-century competition that had begun the sequence. (pp. xx-xxi)

Of course, Shippey does not rely on popularity alone to argue for Tolkien's place of honor among twentieth century authors, but it is not my intent to discuss the literary merit of *The Lord of the Rings* here. Instead, I simply stand before you unashamed to declare my feelings of affection towards J.R.R. Tolkien.

But I began with the plural form, "Tolkiens of My Affection," because I also want to include the author's son, Christopher Tolkien, who has worked as a posthumous editor of his father's work, publishing collections such as *The Silmarillion* (Tolkien, 1977) which provides the history leading up to *The Lord of the Rings*, *Unfinished Tales* (Tolkien, 1980), *The Letters of J.R.R. Tolkien* (Tolkien, 1981), and the 12 volume series, *The History of Middle Earth* (Tolkien, 1983a, 1984, 1985, 1986, 1987, 1988, 1989, 1990, 1992, 1993, 1994, 1996).

The History of Middle Earth is itself an incredible achievement, not a literary one, but in its own way an amazing work of scholarship. For what Christopher Tolkien has done is to go through all of his father's papers and present us with a vast variety of drafts, revisions, and variations of Tolkien's work, much of it related to *The Silmarillion*, about a quarter of it to *The Lord of the Rings*. Tolkien's son provides a painstaking and loving account of his father's writing process, down to the level of reporting to us about the erasures on the page, what has been written over the erasure, and whenever possible what appears to have been erased. He presents a remarkable level of detail that can be quite fascinating in its accounting of the mechanics and materiality of his father's writing. It is sobering to consider that wartime paper shortages affected Tolkien's writing process. And it is uplifting, at least for us academics, to remember that he was grading examinations when he came upon a blank page in a student's exam booklet, was relieved to find one less page to read, and was moved to write on that page the sentence that started it all: "In a hole in the ground there lived a hobbit" (see Shippey, 2000, pp. 1-2).

But overall, the level of detail that Christopher Tolkien provides can, in all honesty, also be quite overwhelming, and even tedious at times. Still I find myself moved, again, by the emotional undercurrent that I can only imagine is at work in these books. I find myself envying the opportunity he had to go through his father's extensive manuscripts, to follow the marks his father made with his own hand by bringing pen and often pencil to paper and, in so doing, to get to know his father's mind in so incredibly intimate a fashion. For writing, as Christine Nystrom (1987) notes, is about nothing so much as it is pure thought, and pure emotion.

Tolkien and McLuhan

In thinking about the sons of famous writers, I cannot help but notice the parallel between Christopher Tolkien and Eric McLuhan, who edited his father Marshall's work in several collections, and who completed his father's culminating work, *Laws of Media* (McLuhan & McLuhan, 1988). And this led me to consider the other connections between Tolkien and McLuhan. For example, both men were Catholics, and both the product of conversion, although for Tolkien it was his mother who made the decision to convert when he was a child. Both were influenced by their religion, but refrained from making overt references to Catholicism in their writing. Both came from the colonies of Great Britain, McLuhan was the consummate Canadian, Tolkien was a son of South Africa although he grew up in England. Tolkien was a product of Oxford, while McLuhan came to England to study at Cambridge. There may even have been some social or scholarly network links between the two; for example, Owen Barfield, one of Tolkien's fellow Inklings, influenced McLuhan's ideas about sense perception. Both McLuhan and Tolkien enjoyed great popular success in the United States, particularly during the sixties, but both failed to gain the acceptance of critics and scholars in their lifetimes. Followers of Tolkien and McLuhan both tend to have strong feelings about the authors, as do their detractors, and consequently there tends to be a sharp dividing line between the two. The process of understanding and appreciating both writers is often described as a kind of gestalt

161

perception, either you "get it" or you don't, and as a kind of epiphany and religious conversion in itself. Moreover, both Tolkien and McLuhan had a particular affinity for the Middle Ages, and a suspicion of modernity. Tolkien, as a philologist, was an expert in languages as they relate to literature and culture. McLuhan began by studying literary theory and criticism, identified himself with the grammar and rhetoric of the medieval trivium (and therefore language and literature; see McLuhan, 2006), and went on to be a leading scholar of culture, communication, and media.

Now, it may be that these connections are simply similar patterns of experience. But I have a particular interest in McLuhan, and in media ecology, the field that he helped to form, as non-Aristotelian approaches closely allied with general semantics (or as Neil Postman put it, "general semantics writ large"—see Moran, 2007/2008; Postman, 1974). And those of us interested in this field sometimes indulge in a form of intellectual play and ask whether a scholar or writer not previously associated with media ecology might, in fact, be a media ecologist. And so I raise the question, is Tolkien a media ecologist? That is, did he have an understanding of media and of symbolic form, an understanding of how they might play a leading role in human affairs? You may already have guessed that the answer is yes, or else I would not have posed the question.

One way to understand media, and media ecology, is by understanding language and the study of languages (see Strate, 2006). This was abundantly clear to Louis Forsdale, the Columbia University professor of English Education who championed McLuhan back in the fifties, and who was a mentor to Neil Postman and many other early media ecology scholars. Forsdale (1981) explained that McLuhan's media ecology was an extension of the linguistic theory associated with Benjamin Lee Whorf, Edward Sapir, Dorothy Lee, and others, the hypothesis that the particular language we speak influences and shapes the way we understand and experience the world; this theory of linguistic relativism is generally considered to be complementary to the general semantics of Alfred Korzybski, S. I. Hayakawa, and Wendell Johnson. During the fifties, McLuhan's colleague, the anthropologist Edmund Carpenter described media as the new languages (Carpenter & McLuhan, 1960), and in *Understanding Media* (2003) McLuhan devoted a chapter to media as translators, as metaphors, and as languages; he also discussed language and speech as forms of media.

It is true that Tolkien did not concern himself very much with media, even in the broad sense that McLuhan employed, nor was he entirely sympathetic to the field of linguistics, that is the scientific study of language. Philology is an older discipline, a humanistic approach to language research that emphasizes the history of language, and its cultural and literary context, what we might otherwise term the pragmatics of communication (Watzlawick, Bavelas, & Jackson, 1967). Significantly, philology is the modern equivalent of grammar in the medieval trivium, the part of the trivium that McLuhan had passionately championed (for him, to truly understand media you had to be a grammarian). Tom Shippey (2000) identifies himself as a proponent of philology within the English curriculum, and in this sense on the same side as Tolkien in battles over academic requirements. Therefore, his insights into Tolkien are of particular relevance for us, and when Shippey declares that "in philology, *literary and linguistic study are indissoluble*" (p. xvii), I can't help but hear echoes of McLuhan's famous maxim, "the medium is the message." For Tolkien, the study of historical works such as the *Elder Edda*, the *Kalevala*, *Beowulf* (which he delivered a significant scholarly lecture on, see Tolkien, 1983b), and *Sir Gawain and the Green Knight* (an edition of which he co-edited, see Tolkien, 1975) was indissoluble from the historical study of languages.

Shippey also places Tolkien in the context of two other twentieth century English authors, James Joyce and George Orwell. McLuhan found in Joyce a major influence and source of inspiration, and he would certainly consider Joyce a media ecologist (see McLuhan, 1969; McLuhan & Fiore, 1968), as would his son, Eric (1997, 1998) and his first graduate student, Donald Theall (1995, 1997). Joyce enjoyed the kind of critical success that eluded Tolkien, but the two held in common a great love of language, and a powerful impulse towards linguistic creativity and play. Orwell also influenced McLuhan, and Postman (1970) included him in his earliest list of media ecologists. And while Orwell's attitude towards language lacked the joy found in Joyce and Tolkien, his Appendix to *1984*, "The Principles of Newspeak," stands as one of the most interesting fictional applications of the Sapir-Whorf-Lee hypothesis, specifically suggesting that controlling language is tantamount to thought control (Orwell, 1949, pp. 246-256). Newspeak is therefore the modern equivalent of the Black Speech of Mordor. Tolkien's fiction shares the view that there is an intimate connection between language, thought, and culture, but as a scholar, his approach is more sophisticated. Shippey (2000) explains some of Tolkien's more interesting views on language:

> He thought that people . . . could detect historical strata in language without knowing how they did it. They knew that names like Ugthorpe and Stainby were Northern without knowing they were Norse; they knew that Winchcombe and Cumrew must be in the West without recognizing that the word *cûm* is Welsh. They could feel linguistic style in words. Along with this, Tolkien believed that languages could be intrinsically attractive or intrinsically repulsive. The Black Speech of Sauron and the orcs is repulsive. When Gandalf uses it in 'The Council of Elrond', 'All trembled, and the Elves stopped their ears'; Elrond rebukes Gandalf for using the language, not for what he says in it. By contrast Tolkien thought that Welsh and Finnish were intrinsically beautiful; he modelled his invented Elf-languages on their phonetic and grammatical patterns . . . It is a sign of these convictions that again and again in *The Lord of the Rings* he has characters speak in these languages *without bothering to translate them*. The point, or a point is made by the sound alone. (p. xiv)

Shippey then goes on to write

> But Tolkien also thought—and this takes us back to the roots of this invention—that philology could take you back even beyond the ancient texts it studied. He believed that it was possible sometimes to feel one's way back from words as they survived in later periods to concepts which had long since vanished, but which had surely existed, or else the word would not exist. (p. xiv)

And thus, Shippey concludes that

> However fanciful Tolkien's creation of Middle-earth was, he did not think that he was *entirely* making it up. He was 'reconstructing', he was harmonizing contradictions in his source-texts, sometimes he was supplying entirely new concepts (like hobbits), but he was also reaching back to an imaginative world which he believed had

once really existed, at least in a collective imagination: and for this he had a very great deal of admittedly scattered evidence. (p. xv)

But *The Lord of the Rings* is not only strongly influenced by Tolkien's scholarly background, it is very much a product of his love of languages. The linguistic medium was Tolkien's message for the very reason that he began by constructing fictional forms of speech, and only after constructing his imaginary tongues did he then go on to create myths and legends as the content of his "Elf Latin" (such as appear in the *Silmarillion*), and still later the novels we know as *The Hobbit* and *The Lord of the Rings*. Tolkien explained in a letter that appeared in the *New York Times* that, "I am a philologist, and all my work is philological" (Shippey, 2000, p. xiii). In a subsequent letter to his American publishers, he went on to elaborate,

> the remark about 'philology' was intended to allude to what is I think a primary 'fact' about my work, that it is all of a piece, and *fundamentally linguistic* in inspiration . . . The invention of languages is the foundation. The 'stories' were made rather to provide a world for the languages than the reverse. To me a name comes first and the story follows. (p. xiii)

Tolkien's creation of fictional languages might be construed to be a thought experiment in philology, or simply a form of scholarly play, but it is also a work of extraordinary imagination. Even the humble origin of the hobbit follows this pattern of formal cause, as first Tolkien writes the sentence on the exam paper, "In a hole in the ground there lived a hobbit," not knowing its meaning. Afterwards, he analyzes the linguistic roots of *hobbit* and realizes it actually means *hole-dweller*. And from this understanding of this one sentence, he then goes on to write all of the remaining sentences that make up the novel known as *The Hobbit*.

The Ecology of the Ents

The medium of language, then, is the hidden ground of *The Lord of the Rings*, one that I believe becomes most visible in *The Two Towers*, in the chapter entitled "Treebeard". It is there that Merry and Pippin meet the Ents, an ancient race of tree-like giants who guard the forests and herd their trees just as humans herd sheep. I have always been fond of this part of the novel, even before I understood its significance, but I must also admit that the encounter with the Ents is, in many ways, unnecessary. It is true that it allows Merry and Pippin to make their first independent contribution to the War of the Ring. And it is true that the Ents attack and defeat Saruman, eliminating the second greatest evil, and freeing the riders of Rohan to come to Gondor's rescue. But the important battle against Saruman was fought at Helm's Deep, and Tolkien does not even describe the actual attack of the Ents on Isengard, merely their march and the aftermath of their victory. They therefore are something of a *deus ex machina*. Moreover, it would have been a simple enough task to describe Isengard as defenseless after Helm's Deep, and have Gandalf dispose of the tower of Orthanc, or to give the role to some other army, say of elves or dwarves.

No wonder, then, that Peter Jackson felt the need to tamper with this part of the story, giving Merry and Pippin a stronger role in the film as the ones who persuade the Ents to go to war, while omitting entirely the extended conversations between the Hobbits and the Ents. This may be attributed to the translation from a verbal medium to an essentially visual one, and in this instance one dominated by special effects (I also think that this is one of a few instances in which Ralph Bakshi's 1978 animated film based on the first half of *The Lord of the Rings* is a better and more faithful adaptation). But it is in the novel alone that we find the full range of Tolkien's philological thought. For example, after Merry and Pippin meet their first Ent, who explains that some call him Fangorn and others call him Treebeard, he goes on to say that he won't give them his true name, explaining

> 'For one thing, it would take a long while: my name is growing all the time, and I've lived a very long, long time; so *my* name is like a story. Real names tell you the story of the things they belong to in my language, in the Old Entish as you might say. It is a lovely language, but it takes a very long time to say anything in it, because we do not say anything in it, unless it is worth taking a long time to say, and to listen to.
>
> 'But now . . . what is going on? What are you doing in it all?

I can see and hear *(and* smell *and* feel) a great deal from this, from this, from this *a-lalla-lalla-rumba-kamanda-lind-or-burúmë.* Excuse me, that is part of my name for it; I do not know what the word is in the outside languages; you know, the thing we are on, where I stand and look out on fine mornings, and think about the Sun and the grass beyond the wood, and the horses, and the clouds, and the unfolding of the world. (pp. 85-86)

A little further on in the chapter, the hobbits suggest the word *hill*, and Treebeard responds by saying, *"Hill?* Yes, that was it. But it is a hasty word for a thing that has stood here ever since this part of the world was shaped" (p. 87). Later still, Treebeard reveals something about the origins of Old Entish, and with it the Ents' position in the conflicts of Middle-earth:

I am not altogether on anybody's *side,* because nobody is altogether on my *side, if* you understand me: nobody cares for the woods as I care for them, not even Elves nowadays. Still, I take more kindly to Elves than to others: it was the Elves that cured us of dumbness long ago, and that was a great gift that cannot be forgotten, though our ways have parted since. (p. 95)

The notion of being cured of dumbness is a curious one, unless we recall that Tolkien was actually a *doctor* of languages.

Apart from language in general, media ecologists are also concerned with the distinction between oral and written language, and with cultures shaped by literacy and cultures characterized by its absence and the presence of oral tradition. Professor Tolkien could hardly be unaware of these issues himself, as he was a contemporary of Milman Parry (1971), who established the nonliterate origins of the Homeric epics and documented the oral composition characteristic of the singers of tales in early twentieth century Serbo-Croatia. The Ents appear to be an oral culture, which is why their names are so long; in reality, the full name of a member of an oral culture might include the recitation of an entire genealogical line of descent. And when the hobbits first meet Treebeard, he is puzzled as to what they are because they do not fit into the categories he memorized in the form of song. Thus, he says,

You do not seem to come in the old lists I learned when I was young. But that was a long, long time ago, and they may have made new lists. Let me see! Let me see! How did it go?

Learn now the lore of Living Creatures!

First name the four, the free peoples:

Eldest of all, the elf-children;

Dwarf the delver; dark are his houses;

Ent the earthborn, old as mountains;

Man the mortal, master of horses. (p. 84)

The solution to Treebeard's uncertainty, offered by the hobbits, fits exactly with what Walter Ong (1982) described as the homeostatic nature of oral cultures. Pippin asks, "Why not make a new line?" and then suggests, "*Half-grown hobbits, the hole-dwellers.*" He then adds, "put us in amongst the four, next to Man (the Big People), and you've got it" (p. 85). Whether this also means changing the number four to five is left up in the air, but after all Tolkien was a philologist, not a mathematician.

The War of the Ring and the Ear vs. the Eye

Media ecologists such as McLuhan and Ong link the distinction between orality and literacy to the distinction between the two primary sense organs, the ear and the eye. They describe a kind of war between the two over the course of world history, as the ear dominates for most of human history, but the eye gains ascendancy in ancient Greece and Rome, and again in modern Europe and America (until finally overcome by the electronic media's retrieval of acoustic space and secondary orality). This is not a theme that Tolkien emphasizes, and yet he does tend to favor the acoustic over the visual in his treatment of good vs. evil. The Ring is itself an object of visual beauty, associated with possessiveness and greed, and a magic item that acts on the visual sense, both in rendering its user invisible, but at the same time more visible to the Black Riders and Sauron than would otherwise be the case. For example, at the end of *The Fellowship of the Ring*, in "The Breaking of the Fellowship," Frodo slips on his ring to escape Boromir, and finds himself gifted with a vision of Middle-earth at war that only induces despair in him. Then, Tolkien (1965a) writes

> And suddenly he felt the Eye. There was an eye in the Dark Tower that did not sleep. He knew that it had become aware of his gaze. A fierce eager will was there. It leapt towards him; almost like a finger he felt it, searching for him. Very soon it would nail him down, know just exactly where he was. . . .
>
> He heard himself crying out: *Never, never!* Or was it: *Verily I come, I come to you?* He could not tell. Then as a flash from some other point of power there came to his mind another thought: *Take it off! Take it off! Fool, take it off! Take off the Ring!*
>
> The two powers strove in him. For a moment, perfectly balanced between their piercing points, he writhed, tormented. Suddenly he was aware of himself again. Frodo, neither the Voice nor the Eye: free to choose, and with one remaining instant in which to do so. He took the Ring off his finger. (p. 519)

The Voice was, of course, Gandalf, who had fallen in Moria, but returned to oppose the Eye of Sauron. And Gandalf opposed the White Hand of Saruman, the image of the hand symbolizing Saruman's identity as a technologist. Moreover, Sauron's Ring is a wonderful example of what McLuhan calls the extensions of man, a medium or technology that extends Sauron's reach but amputates (literally) his power when Isildur cuts off both ring and finger. Even the technologies of the free peoples present a Faustian bargain, as Neil Postman (1992) would put it. For example the Palantíri or looking stones of Aragorn's ancestors, Middle-earth's telecommunications system, became a means by which Sauron could gain access to others and corrupt them, as he did with Saruman and Denethor, Steward of Gondor; this sounds an awful lot like television as described by Postman (1985). And the three jewels known as the Silmarils, forged by the Elf Fëanor, become the source of great conflict among the Elves and the higher powers. The light they contain is good and beautiful, but in capturing and containing the light, Fëanor sets the stage for much evil.

In the war between the ear and the eye, Tolkien, like most media ecologists, is on the side of the ear. Thus, sound is prioritized in his fictional account of the beginning of the world, which serves as the first chapter of the *Silmarillion* and is titled "The Music of the Ainur". It starts with

> There was Eru, the One, who in Arda is called Ilúvatar; and he made first the Ainur, the Holy Ones, that were the offspring of his thought, and they were with him before aught else was made. And he spoke to them, propounding to them themes of music; and they sang before him, and he was glad. (Tolkien, 1977, p. 3)

The Ainur are angels, although in some respects they more closely resemble the gods of Greek and Norse mythology. At first they are unable to join together in unison to create music, but then Ilúvatar acts:

> Then Ilúvatar said to them: 'Of the theme that I have declared to you, I will now that ye make in harmony together a Great Music. And since I have kindled you with the Flame Imperishable, ye shall show forth your powers in adorning this theme, each with his own thoughts and devices, if he will. But I will sit and hearken, and be glad that through you great beauty has been wakened into song.'
> Then the voices of the Ainur, like unto harps and lutes, and pipes and trumpets, and viols and organs, and like unto countless choirs singing with words, began to fashion the theme of Ilúvatar to a great music; and a sound arose of endless interchanging melodies woven in harmony that passed beyond hearing into the depths and heights, and the places of the dwellings of Ilúvatar were filled to overflowing, and the music and echo of the music went out into the Void, and it was not void. (pp. 3-4)

As Tolkien continues on with his creation myth, the greatest of the Ainur, Melkor, takes on the role of Lucifer in challenging Ilúvatar's theme, and weaving in discordant notes of his own devising. Ilúvatar asserts himself, and although strife has been introduced in the midst of harmony, he tells Melkor and the rest that "no theme may be played that hath not its uttermost source in me, nor can any alter the music in my despite. For he that attempteth this shall prove but mine instrument in the devising of things more wonderful, which he himself hath not imagined" (p. 4). The Ainur are put in their place, at which point we finally move from the ear to the eye:

Ilúvatar said to them: 'Behold your Music!' And he showed to them a vision, giving to them sight where before was only hearing; and they saw a new World made visible before them, and it was globed amid the Void, and it was sustained therein, but was not of it. And as they looked and wondered this World began to unfold its history, and it seemed to them that it lived and grew. (p. 6)

Music is followed by vision here, just as Genesis begins with God's speech act, "Let there be light," and then continues with the actual appearance of light. Tolkien's creation parallels Genesis in certain ways. As a symbolic form, it conveys the *feeling* of Genesis, and of Western creation myths in general. But it is not consistent with the Biblical account of creation. It is religious, but not specifically Catholic or in the Judeo-Christian tradition. Along the same lines, overt references to Christianity, such as those that appear in the *Perelandra* trilogy written by his fellow Inkling, C. S. Lewis, are not present in Tolkien's fiction. There are familiar motifs, for example Gandalf's death and resurrection, and Frodo's self-sacrifice and essential crucifixion—hence the nineteen sixties slogan "Frodo Lives!" But these elements do not by any means add up to an allegory. Moreover, Tolkien wrote in the Foreword to the second edition of *The Lord of the Rings*, "I cordially dislike allegory in all its manifestations, and always have done so since I grew old and wary enough to detect its presence. I much prefer history, true or feigned, with its varied applicability to the thought and experience of readers" (Tolkien, 1965a, p. xi).

As a philologist, Tolkien was an historian as well as a linguist and literary theorist. And his fiction reflects the historical consciousness that emerged in the nineteenth century, and is reflected in the philosophy of Hegel and Marx, the world histories of Toynbee and Spengler, the media ecology of Lewis Mumford and Harold Innis, and the nonfiction and science fiction of H. G. Wells and Isaac Asimov. Moreover, the combination of language and history forms the basis of something more than a narrative. It provides us with an environment, a symbolic environment, a media environment. We become immersed in Tolkien's world, and as in the baptismal ritual, immersion leads to conversion. Today, it has become commonplace to talk about universes, the DC and Marvel Universes in comic books, the *Star Trek* and *Star Wars* universes in television and film, and game playing universes such as was pioneered under the name

Dungeons and Dragons. Tolkien's act of creation gave us the first of these fictional universes, and remains the model that others still draw upon today.

Tolkien's historical consciousness extends to form as well as content, as *The Lord of the Rings* incorporates elements of the medieval manuscript. He breaks up the otherwise homogenous text of the novel with poetry and song, some of which is in English, some in Elvish without translation. He mixes into the narrative bit and pieces of his fictional myths, legends, and histories. And he includes illustrations with Elvish writing, maps, genealogical tables, and other appendices. In this, we can draw a parallel to McLuhan's untraditional books, notably the *Gutenberg Galaxy* (1962) which also retrieved elements of scribal manuscript production. In both instances, critics had difficulty understanding the formal innovations as well as the highly original content that these works contained.

By way of conclusion, I would suggest to you that the power of Tolkien's work, his ability to elicit such strong emotion in his readers, has much to do with the fact that he was a media ecologist, that he understood the media of speech, language and symbolic form. And for myself, understanding media has helped me in understanding Tolkien, and served to deepen my affection for the author and his works.

Chapter Ten
Poetry Ring

Wendell Johnson (1946) said that "the better part of science is the language of science" (p. 50) and like science, poetry has a language all its own. We therefore might consider the corresponding claim, that the better part of poetry is the language of poetry. Certainly, following the lead of the brilliant philosopher of symbolic form, Susanne K. Langer (1953, 1957), we can understand that the language of poetry can be used to express and represent thoughts, feelings, and perceptions that the language of science cannot. We can understand as well that the language of poetry can evoke aspects of human experience that no other language may be able to grasp, or even touch. General semantics, being truly *general*, is concerned with all forms of meaning making, all manner of symbolic communication. Moreover, unlocking the creative potential within individuals has also been a longstanding focus of this non-Aristotelian system. Poetry was an oral medium in its origins, a form of speech, and song (see, for example, Havelock, 1963; Ong, 1982). As such, poetry originally served a mnemonic function, it was a means of preserving knowledge, and therefore can be understood as the most basic means by which the time-binding capacity of language was extended, prior to the invention of writing.

II

Alfred Korzybski, in his discussion of higher order abstractions in *Science and Sanity* (1993), notes that the terms that are most central to our lives, such as *truth*, *love*, and *meaning*, terms that are also quite central to much of modern poetry, are highly ambiguous words that are used on many different levels or orders of abstraction. Korzybski refers to this property of language as *multiordinality*, which he abbreviates as *m.o.*, and argues that one of the benefits of becoming conscious of the multiordinality of language is that "the whole linguistic process becomes extremely flexible, yet it preserves its essential extensional one-valued character, in a given case" (p. 437), which is to say that we can assign specific definitions as needed, without reifying terms, while at the same time enjoying the aesthetic pleasures of ambiguity. Given his emphasis on scientific method and rationality, it would be easy to assume that Korzybski only valued the languages of science and mathematics, and had no room for the arts. But quite to the contrary, he had a great appreciation for the arts, in fact was married to the noted American painter, Mira Edgerly; and as for poetry, he wrote the following in *Science and Sanity* as part of his discussion of multiordinality:

> In a certain sense, such a use of *m.o.* terms is to be found in poetry, and it is well known that many scientists, particularly the creative ones, like poetry. Moreover, poetry conveys in a few sentences more of lasting values than a whole volume of scientific analysis. The free use of *m.o.* terms without the bother of a structurally impossible formalism outside of mathematics accomplishes this, *provided we are conscious of abstracting; otherwise only confusion results.* (p. 437; emphasis in the original)

Korzybski, then, was not only non-Aristotelian, but non-Platonic, insofar as Plato thought to ban poetry from his ideal society.

III

General semanticists emphasize the role that the languages of science, logic, and mathematics play in the process of abstracting. We also acknowledge that the arts constitute a means of abstracting from reality, albeit in a different manner. Similarly, Camille Paglia argues in *Sexual Personae* (1990)

> I believe that the aesthetic sense . . . is a swerve from the chthonian. It is a displacement from one area of reality to another, analogous to the shift from earth-cult to sky-cult. Ferenczi speaks of the replacement of the animal nose by human eye, because of our upright stance. The eye is peremptory in its judgments. It decides what to see and why. Each of our glances is as much exclusion as inclusion. We select, editorialize, and enhance. Our idea of the pretty is a limited notion that cannot possibly apply to earth's metamorphic underworld, a cataclysmic realm of chthonian violence. We choose not to see this violence on our daily strolls. Every time we say nature is beautiful, we are saying a prayer, fingering our worry beads. (p. 15)

As methods of abstracting, the languages of the arts represent a means of imposing order on chaos, constructing maps that provide us with the illusion of a stable, predictable, and navigable territory. As Paglia explains

> Art makes *things*. There are, I said, no objects in nature, only the grueling erosion of natural force, flecking, dilapidating, grinding down, reducing all matter to fluid, the thick primal soup from which new forms bob, gasping for life. Dionysus was identified with liquids—blood, sap, milk, wine. The Dionysian is nature's chthonian fluidity. Apollo, on the other hand, gives form and shape, marking off one being from another. All artifacts are Apollonian. Melting and union are Dionysian; separation and individuation, Apollonian. . . . Every artist who is compelled toward art, who needs to make words or pictures as others need to breathe, is using the Apollonian to defeat chthonian nature. (p. 30)

The metaphor of Dionysus corresponds to a movement towards lower levels of abstraction, and that of Apollo to higher ones. But we do not use art only to impose order and organization on our experience of a chaotic world; we can also utilize one of the arts to describe or comment upon another art form. The term *ekphrasis* refers to any instance in which one work of art addresses or takes as its content another artwork. To an extent, then, we can liken ekphrasis in the languages of art to the process of translation that we associate with the languages of speech and writing.

Ekphrasis can be seen as Apollonian in one sense, a move to a higher level of abstraction, as for example a sculpture may be said to represent a particular individual, and a painting of the sculpture then represents the sculpture that represents the individual. But ekphrasis also summons a touch of the Dionysian, as it crosses the boundaries between one form and another, and therefore somewhat muddies their distinctions. We might then think of it as not so much a move to a higher or lower level of abstraction as a lateral move across modes of abstraction.

IV

As far as the laws of mathematics refer to reality, they are not certain; and as far as they are certain, they do not refer to reality.
–Albert Einstein, *Sidelights on Relativity* (1983, p. 12)

177

On the Binding Biases of Time

As a deductive system, mathematics is rooted in the Aristotelian logic of identity, noncontradiction, and the excluded middle. As a means of measuring the physical world, however, mathematics makes it possible for us to construct some of our most precise maps of reality. And as a symbol system, mathematics can be distinguished from other languages by its lack of ambiguity. As Stendhal (2002) states in *The Life of Henri Brulard*, "I loved, and still do love, mathematics for itself as not allowing room for hypocrisy and vagueness" (p. 120). As such, the language of mathematics holds out the hope for universal communication, and unanimity, as a language through which and within which human beings may interact with complete clarity and honesty, and therefore without issue or conflict. And as Bertrand Russell noted, "The most savage controversies are those about matters as to which there is no good evidence either way. Persecution is used in theology, not in arithmetic" (p. 22). This may, for the most part, be true, however, George Orwell's equation $2+2=5$, as expressed in his novel *1984* (1949), is an example of how mathematics may indeed be used to persecute; Orwell's fictional example of arithmetical doublethink, it is important to add, was derived from the reality of Stalin's declaration that a 5-year plan could be completed in 4 years.

At first glance, the language of mathematics and the language of poetry may appear to be diametrically opposed to one another, poetry being a language that makes frequent use of ambiguity, contradiction, and even non-identity in an intuitive fashion. Mathematics is the most digital and discursive of codes, while poetry pushes human language into the analogical and presentational realm of the arts, into the realm of music, painting, sculpture, and dance. And yet, Albert Einstein (1935) noted that "pure mathematics is, in its way, the poetry of logical ideas" (p. 12). Moreover, Thomas Hill (1857) stated that mathematics is "usually considered as being the very antipodes of Poesy. Yet Mathesis and Poesy are of the closest kindred, for they are both works of the imagination" (p. 229). It is not surprising, then, to find that poets have utilized the language of mathematics in the manufacturing of their poems. Everyone has heard metaphorical statements related to physics such as: the *power* of love, the *force* of oppression, or the *gravity* of the situation, and mathematics too may offer ample opportunities for metaphoric transpositions. The fusing of the language of math together with verbal art offers a wonderfully rich combination of aesthetics.

Poets take the language of mathematics and may use it directly or may contort it. They may twist the functions of mathematical terms in an attempt to convey their own personal meanings. Poets will take the neatly defined and seemingly solid language of mathematics and pull it out, extend it, expand it and distort it into something quite unexpected. In doing so, the poet calls our attention to mathematics as a human invention, as a set of symbols, as a language, rather than the more common representation of mathematics as a reflection of reality or truth.

V

Think of general semantics, and in all likelihood the first thing to come to mind will be Alfred Korzybski's famous aphorism, *the map is not the territory*. The map is a metaphor, of course, for words, symbols, and the entire process of abstracting, while territory stands in for things and events, otherwise known as reality. As human beings, we create maps based on our experiences, maps made out of words, images, and other types of symbolic form. And we are guided by the maps we make, so much so that we take them for granted and forget that they are only maps of our own devising and nothing more, that they are necessarily subjective and inevitably inaccurate in some way, most definitely incomplete, and almost certainly out of date.

Poetry too is a map of sorts, and in its ancient mnemonic function, could serve as a memorable means of recalling a set of verbal directions. Modern poets more typically use their art to make maps of the inner world of thought, sensation, and emotion. The tension between the intensional orientation of poetry and the extensional orientation of mapmaking also makes for interesting effects. The map is generally perceived as a product of science, mathematics, and engineering, one that is evaluated in regard to its accuracy, the result being that we blur the line between map and territory. The poetic perspective on maps emphasizes maps as *maps*, as objects of contemplation in and of themselves, rather than as transparent windows onto reality; the poetic perspective serves to remind us that maps also have a subjective side, eliciting fascination and wonder, curiosity and mystery, a sense of place and a sense of displacement both, and again, a metaphor for ourselves, our actions and our thoughts.

VI

Few topics are more central to the discipline of general semantics than language, specifically what language does for us, and what language does to us. And it is a commonplace to say that language, and more generally the capacity for symbolic communication, is the defining characteristic of the human species. In his classic work on general semantics, *Language Habits in Human Affairs* (1994), Irving J. Lee argues, "to be concerned with language as used by living people is to bring us to the heart of things *human*" (p. 3). Acknowledging the close relationship between language as a symbol system and speech as its primary manifestation, he goes on to note that, "without language, written and spoken, the silence of the day would be broken only by shadowy forms, primitive cries and grunts, the sounds of the winds and the waves, the rustle and murmur of moving things" (p. 3). And later, he invokes the concept of time-binding that is foundational to general semantics as a system, the idea that language enables us to accumulate knowledge, pass it on from generation to generation, eliminate error in the process, and make progress as a society: "To see the uniqueness of man's time-binding capacity is to begin to realize the significance of language. If we discover the creative uses of words, we may begin to know what it is to function *humanly*" (p. 5).

It follows that poetry, insofar as it represents one of the most creative ways in which we use words, is a pathway to knowing what it means to be human, and to function in a human manner. Poetry has its origins in memory and mnemonics, and therefore carries an intimate connection to time-binding, the unique characteristic of the human race as a class of life. And poetry opens the door to language for the sake of language, an emphasis on the form of language as opposed to its content. As such, it can make us conscious of language, make us pay attention to language as a mode of communication and a mirror reflecting ourselves, rather than as a transparent window onto "reality". Poetry has the potential to raise our awareness of the unique properties and biases of language as a medium or symbol system, to function as language about language, as a *metalanguage*.

VII

Language, as Korzybski (1993) made clear, can be self-reflexive. We can use language to discuss language, and then we can use language to discuss our language about language. This process can continue to escalate with no end in sight. We can make statements about statements, statements about statements about statements, and so on. We can ask questions about questions, and questions about questions about questions, in infinite recursion. If our words are meant to be mirrors of reality, they can also mirror themselves in much the same way as when two mirrors are held opposite to one another.

Poetry too is subject to the principle of self-reflexiveness. Poets by the very nature of their craft, live in the world of words, and use words to make reference to that world. On one level, poets attempt to appeal to the senses and convey their observations and reflections about the world, through their poetry. In recognizing that emotions and feelings are not separate from thought, poets seek to select language that will effectively and compellingly convey the more subtle extensions of their perceptions and experience.

Readers bring with them their own personal and projective abstractions and this leads to varying receptions and interpretations of the poem. This is one challenge that provokes poets to pay close attention to their language choices. When choosing the specific language to use in their poetry, they engage in an active process of seeking and selecting words, and this will often lead them to reflect on the nature of language and indeed the very action of writing itself. In doing so, poets may be led to create self-reflexive poetry, that is to say, poetry that examines language, the poem and the act of writing itself. In this way, poetry may function as a probe that can help us to better understand language and symbolic form, as McLuhan (1962, 1969, 2003) has noted.

When written well, self-reflexive poetry will be personally clearing for poets, allowing them to explore and examine why they write, why they may be having difficulty writing (perhaps even alleviating writer's block) and the limitations of words themselves. Further to this, poets may write poems about the poetry of self-reflexiveness, that is, poetry about poetry about poetry, and this process could continue unabated, limited only by the imagination of poets and their audience.

VIII

General semanticists by and large are well acquainted with the story of how a young engineer from Poland named Alfred Korzybski was horrified by his experiences during the First World War, the first war to use weapons of mass destruction, the war that put an end to the notion that war is glamorous, glorious, an occasion for celebration rather than a tragedy. And motivated by what he had witnessed, Korzybski asked, why can't we build a peaceful world for ourselves? Why can't we make the same kind of progress in human relations and ethical conduct as scientists and engineers make in their endeavors? Simply put, why can't humanity grow up and settle its conflicts like mature adults? Korzybski posed these questions in his first book, *Manhood of Humanity* (1950), where he characterized our species as a time-binding class of life, that is, as a species capable of storing and passing knowledge down from generation to generation. In turn, he answered them in his second book, *Science and Sanity* (1993), where he introduced the non-Aristotelian discipline of general semantics. Consequently, over the past three quarters of a century, general semantics has been offered as an antidote to the insanity of war and conflict.

War is also a topic for poetry, one perhaps as old as poetry itself. For example, the Trojan War was the subject of Homer's *Iliad*, which begins with the wrath of Achilles and ends with the funeral of Hector. Certainly, one of the best known poems in the English language is the nineteenth century "The Charge of the Light Brigade" by Alfred, Lord Tennyson. Celebration gave way to critique during the twentieth century, paralleling social and political developments. One of the first to engage in antiwar poetry was e e cummings who, like Korzybski, was profoundly disturbed by what he witnessed during World War I, and wrote the following lines in 1926:

> my sweet old etcetera
>
> aunt lucy during the recent
>
> war could and what
>
> is more did tell you just

what everybody was fighting

for,

my sister

isabel created hundreds

(and

hundreds)of socks not to

mention shirts fleaproof earwarmers

etcetera wristers etcetera,my

mother hoped that

i would die etcetera

bravely of course my father used

to become hoarse talking about how it was

a privilege and if only he

could meanwhile my

self etcetera lay quietly

in the deep mud et

cetera

(dreaming,

et

cetera,of

Your smile

eyes knees and of your Etcetera)

Looking back to the generation that produced Korzybski and Cummings, and across a century in which the world has never been entirely at peace, thinking of the killing fields of the Great War, and the victims and the survivors, and to all those who followed after, we need to be reminded that we all have important work to do, on behalf of all of humanity.

IX

In *Language in Thought and Action* (Hayakawa & Hayakawa, 1990), the most popular and best selling of all books on general semantics, S. I. Hayakawa devotes an entire chapter to "Poetry and Advertising" (chap. 13, pp. 134-143), an odd couple if every there was one. And while acknowledging the differences between the two forms of expression, Hayakawa also notes that "both make extensive use of rhyme and rhythm," "both use words chosen for their affective and connotative values rather than their denotative content," both make ample use of ambiguity and multiple levels and layers of meaning, and "both strive to give meaning to the data of everyday experience; they both strive to make the objects of experience symbolic of something beyond themselves" (p. 135). He goes on to argue that both actively encourage identification (traditionally known as mimesis), and both are creative in that they "have the common function of entering into our imaginations and shaping those idealizations of ourselves that determine, in large measure, our conduct" (p. 137). Hayakawa then proposes

> Let us call this use of verbal magic (or verbal skullduggery) for the purpose of giving an imaginative, symbolic, or ideal dimension to life, and all that is in it, *poetry*. If we speak separately of what we ordinarily call poetry and advertising, let us speak of the latter as *sponsored poetry*, and of the former as *unsponsored poetry*. (p. 137)

Working with this distinction, Hayakawa later acknowledges the difficulties that unsponsored poets face in contemporary American culture:

> The unsponsored poet of today works in a semantic environment in which almost all the poetry that ordinary people hear and read is the sponsored poetry of consumer goods. Poetic language is used so constantly and relentlessly for the purposes of salesmanship that it has become almost impossible to say anything with enthusiasm or joy or conviction without running into the danger of sounding as if you were selling something. (p. 140)

Both advertisers and poets must work with the symbols of their culture, but poets must contend with the fact that advertisers have appropriated most of the significant symbols of our culture. Hayakawa believes that this may explain why "the verse of unsponsored poets is often difficult to understand and full of obscure symbolism" (p. 141). He notes, sympathetically that "they are practically driven to use obscure symbols out of the Upanishads or Zen Buddhism in their search for something the advertisers have not already used" (p. 142). While advertising use the symbols of everyday life to sell us on new products and services, "poets, by creating new ways of feeling and perceiving, help to create the new ways of thinking that bring us to terms with a changing world" (p. 142). This is a vital, invaluable function, one well worth acknowledging, and celebrating. And Hayakawa's conclusion to this chapter is well worth repeating:

> With what symbols shall the poet bring us to terms with the realities of our own times? In the past few decades, whole new areas of thought and exploration have been opened up by the sciences—by electronics, by astrophysics, by microbiology, by the study of nucleoproteins and their role in genetics, by radioactive tracer studies, and by nuclear physics. Instant communications bring us unsettling news from parts of the world that we had never thought about before. Astronauts shoot through space, so that the limits of the planet we live on are no longer the limits of our exploration. We can, and do, describe these new developments in the language of science, but how are we to take these new and urgent realities into our hearts as well as our minds, unless poets give us new images with which to experience them? (pp. 142-143)

Hayakawa leaves us with a question that truly goes to the heart of the matter, and in a fashion that could arguably be considered, in its own right, poetic.

X

Poetry precedes the invention of writing, and there is a world of difference between poetry that is a product of oral composition within an oral tradition, and poetry that is composed with the aid of writing. Of course, there also are significant differences between poetry appearing in print and spoken word poetry, between reading and reciting poetry, and more so listening to its recitation. But the reciting of poetry is still a re-siting, a re-situating of what is primarily a literary form, which means that it is also a re-sighting, as literary forms are governed by the eye in addition to (and sometimes instead of) the ear. As a literary form, then, concern has often been directed to the writing process as it pertains to poetry, and to the process of reading, and interpreting, the poetic form. The literary scholar I.A. Richards, whose perspective on language and meaning was consistent with general semantics, devoted his 1929 study, *Practical Criticism*, to the problem of how individuals read and understand (or misunderstand) poetry. The problem, in part, has to do with the multiplicity of meaning:

> The all important fact for the study of literature—or any other mode of communication—is that there are several kinds of meaning. Whether we know and intend it or not, we are all jugglers when we converse, keeping the billiard-balls in the air while we balance the cue on our nose. Whether we are active, as in speech or writing, or passive, as readers or listeners, the Total Meaning we are engaged with is, almost always, a blend, a combination of several contributory meanings of different types. Language—and pre-eminently language as it is used in poetry—has not one but several tasks to perform simultaneously, and we shall misconceive most of the difficulties of criticism unless we understand this point and take note of the differences between these functions. (p. 174)

The four functions Richards refers to are Sense (literal meaning), Feeling (emotional connotations), Tone (the source's attitude towards and relationship with the receiver), and Intention (the communicator's conscious or unconscious motives). But given that poetry poses special problems for the isomorphic sharing of meaning, a fact borne out by the research on reading reported in *Practical Criticism*, and that these problems are nowhere near as acute for other literary or linguistic forms, we might well ask what is the purpose of poetry? Certainly, at one time poetry was a mnemonic medium, a means of augmenting the storage capacity of the brain and the time-binding ability of a society, a method for preserving knowledge by placing it in a memorable form. But that time has long since past, and poetry has long since evolved from its roots as a form of mnemotechny employed by singers of tales. Perhaps we ought to retain the basic concept of poetry as language given memorable shape and form, but that does not seem sufficient for a modern understanding of poetry (let alone a postmodern one). One answer was supplied by Owen Barfield in a work originally published a year before *Practical Criticism*, entitled *Poetic Diction* (1973):

> An introspective analysis of my experience obliges me to say that appreciation of poetry involves a 'felt change of consciousness'. The phrase must be taken with some exactness. Appreciation takes place at the actual moment of change. It is not simply that the poet enables me to see with his eyes, and so to apprehend a larger and fuller world. He may indeed do this, as we shall see later; but the actual moment of the pleasure of appreciation depends upon something rarer and more transitory. It depends on the change itself. If I pass a coil of wire between the poles of a magnet, I generate in it an electric current—but I only do so while the coil is positively moving across the lines of force. I may leave the coil at rest between the two poles and in such a position that it is thoroughly permeated by the magnetic field; but in that case no current will flow along the conductor. Current only flows when I am actually bringing the coil in or taking it away again. So it is with the poetic mood, which like the dreams to which it has so often been compared, is kindled by the passage from one plane of consciousness to another. It lives during the moment of transition and then dies, and if it is to be repeated, some means must be found of renewing the transition itself. (p. 52)

For the individual, then, poetry is an instrument of perception, but more importantly, it is a way of obtaining education and enlightenment, a means of altering and raising one's consciousness. Beyond what it has to offer the individual, in the conclusion to *Practical Criticism*, Richards (1929) provides a cogent argument for the value of poetry to society as a whole, and why we need poetry now more than ever (an argument that rings even more true today than it did in 1929):

> It is arguable that mechanical inventions, with their social effects, and a too sudden infusion of indigestible ideas,

are disturbing throughout the world the whole order of human mentality, that our minds are, as it were, becoming of an inferior shape—thin, brittle and patchy, rather than controllable and coherent. It is possible that the burden of information and consciousness that a growing mind has now to carry may be too much for its natural strength. If it is not too much already, it may soon become so, for the situation is likely to grow worse before it is better. Therefore, if there be any means by which we may artificially strengthen our minds' capacity to order themselves, we must avail ourselves of them. And of all possible means, Poetry, the unique, linguistic instrument by which our minds have ordered their thoughts, emotions, desires . . . in the past, seems to be the most serviceable. It may well be a matter of some urgency for us, in the interests of our standard of civilisation, to make this highest form of language more accessible. From the beginning civilisation has been dependent upon speech, for words are our chief link with the past and with one another and the channel of our spiritual inheritance. As the other vehicles of tradition, the family and the community, for example, are dissolved, we are forced more and more to rely upon language. (p. 301)

At a time when technology is increasingly calling our very humanity into question, we would do well to remember that being human is bound up inextricably with language, as it enables us to transcend time and space, and that poetry as the highest form of language, is the medium that makes us the most human of all.

ELE

Chapter Eleven

Be(a)Very Afraid

A re you afraid of the word *beaver?* Would you be horrified if someone called you an *eager beaver?* What would you think if you heard someone with a British accent make reference to a person *beavering about?* Honestly, when you hear or read the word *beaver*, what comes to mind? Is it the cute, furry animal known for building dams? Is it the sitcom from the golden age of television, *Leave It to Beaver?* Is it the inexplicably named Samuel J. Gopher character from Disney's animated adaptation of A. A. Milne's *Winnie the Pooh* (and let's leave the topic of *pooh* for another occasion). Or might it be the popular 1990s cartoon series seen on cable television's Nickelodeon channel, *The Angry Beavers?* Or do you have a dirty mind?

The reason I bring this up is a news story out of Canada, which UPI (2010) gave the headline of, "Web Forces Beaver Magazine Name Change." The report notes that the publishers of the periodical that was known for 90 years as *Beaver* magazine decided to change its name to *Canada's History* "because of online pornography and spam filters." You might consider this to be a case of internet scapegoating, but it is not without precedent. Back in 2001, when spam filters were not as prevalent as they are today, an American institution of higher learning, Beaver College, changed its name to Arcadia University. Let me quote from the history section of the Wikipedia entry[*] on this school: "The school was founded in Beaver, Pennsylvania in 1853 as **Beaver Female Seminary** [boldface in original]" (Arcadia College, n.d.). Yes, that's what it says. Why? What were you thinking? And yes, there is a town, actually a borough of Beaver in Pennsylvania, situated on the Beaver River, and located in Beaver County. In fact, it's the county seat, and there's nothing funny about that either. Anyway, the wiki entry goes on to note that, "by 1872 it had attained collegiate status, under the auspices of the Methodist Episcopal Church, and was named **Beaver College** [boldface in original again]."

So, what about the more recent name change? Well, read on:

In July 2001, upon attaining university status, Beaver College

---

[*] It is, of course, understood that Wikipedia entries are subject to continual revision, and therefore difficult to verify. The quotes from this entry were retrieved on January 27, 2010.

officially changed its name to Arcadia University. It was thought that a new name would emphasize the school's position as one of the top small institutions of higher learning on the East Coast, and would cement its change in designation from "college" to "university." The decision was also made in part to shed its association with the former commonly derided name. As then-president Bette Landman noted, "[The name] too often elicits ridicule in the form of derogatory remarks pertaining to the rodent, the TV show *Leave It to Beaver* and the vulgar reference to the female anatomy."

Bette's frank explanation was included in a news story written by Ron Todt (2000), archived at ABCnews.com, and entitled "Beaver College Announces New Name." Interestingly, the article says that "the decision was announced just after midnight at a surprise pajama party for students, who were rounded up from residence halls with less than an hour's notice." No comment on that, please. And after quoting President Landman, Todt went on to note

> Beaver College has appeared on David Letterman's Top 10 list. Conan O'Brien and Howard Stern have made jokes about it. And when *Saturday Night Live* writers invented an annoying film critic for a recent sketch, they made him a representative of Beaver College campus radio.

> The college's own research shows the school appeals to 30 percent fewer prospective students solely because of the name. And the problems worsened with the rise of the Internet, since some Web filters intended to screen out sexually explicit material blocked access to the Beaver College Web site. (Todt, 2000)

So, maybe we should be applauding *Beaver* magazine for holding out as long as it did. Returning to the UPI (2010) story, Deborah Morrison, President of Canada's National History Society and the publisher of the history periodical, was quoted as saying,

> "To be perfectly blunt about it, 'The Beaver' was an impediment on the Internet," she said. "Unfortunately, sometimes words take on an identity that wasn't intended in 1920, when it was all about the fur trade."

Morrison said research showed women and people under the age of 45 were most likely to dislike the magazine's old name and others were complaining their e-mail spam filters were blocking *The Beaver*'s e-newsletters.

It is the second-oldest magazine in Canada after *MacLean's*, a weekly news magazine founded in 1905, the news agency said.

This topic is really quite tricky, if you think about it. After all, I am a man writing about two women presidents presiding over two beaverish name changes, at a time when just saying the word *beaver* in the workplace can have you brought up on sexual harassment charges (I kid you not!), even without making any specific reference to the connotations made famous by pornographer Larry Flynt and *Hustler* magazine. But, all I really meant to do was to call your attention to a blog post by Colby Cosh, a Canadian writer, and an assistant editor at Canada's oldest magazine, *Macleans*. The piece is entitled "Le Castor Fait Tout," which according to my very rusty French means something like *the beaver does it all*, or *the beaver makes everything*, or *the beaver entirely*, or something like that. The name is not important, of course, what is important is what Cosh (2010) has to say:

> I wonder if any of the other Macbloggers have been straining at their imaginations trying to find a PG-rated way to talk about the name change over at Canada's second-oldest magazine. It took me a while to remember that General Semantics has an answer for this. So: *The Beaver,* now to become *Canada's History*, was named in 1920 for what we'll call beaver$_1$, the rodent *Castor canadensis*. The periodical was obliged to make the change because of jokes about and search-engine confusion with beaver$_2$, a colloquialism for an anatomical neighbourhood in the human female.

Aha, you may be saying to yourself, finally we get to the point, a reference to general semantics. And yes indeed, Colby insightfully draws upon the non-Aristotelian principle of non-identity, and he brings into play the extensional device called indexing, which reminds us that words can have many different meanings, can refer to many different "things" or phenomena or events, and can also operate on different levels

193

of abstraction (e.g., *the beaver* pointing to one individual vs. *the beaver* referring to the entire species). Indexing is a device that is meant to remind us that different meanings of a word should not be equated with one another, that beaver$_1$ is not beaver$_2$. And it is meant to remind us to be more precise with our language than we otherwise would be. Of course, indexing is not meant to replace the way we talk or write, but rather serves as an exercise to help us think more clearly and critically, much as it works here for Cosh (2010):

> Beaver$_1$ is, for my money, the best of the popular symbols of Canadian nationhood. Unlike the maple leaf, which is utterly unknown west of Lake of the Woods, the beaver$_1$ is ubiquitous throughout Canada, yet is integral to our history as seen from most regional standpoints. (I was well into my twenties before I saw an actual maple leaf. I had always assumed that the silhouette on the flag was heavily stylized; when I saw just how similar the real thing was to the icon, I almost fell down laughing—partly at my own ignorance.) Perhaps the beaver$_1$ and the maple leaf are best understood as representing "Hamiltonian" and "Jeffersonian" aspects of the Empire of the St. Lawrence, the one signifier standing for the agrarian past of our dreams—hot sweets, home cooking, big families pounding taps into trees at sugaring-off time—and the other representing progress, engineering, and the industrial virtues.

Now, I find this rather fascinating, as I had no idea that the maple leaf was a symbol of the eastern elite. Canada has east-west issues not unlike those of the United States, and the question of national symbol also rings a (liberty?) bell—Benjamin Franklin advocated for the turkey, a bird only found in the New World, over the eagle, as the USA's totem. Of course, that decision was made before the end of the eighteenth century, whereas Canada did not adopt the maple leaf flag until 1965. And while the US does not have any flags with turkeys on them, there is in fact a beaver flag of sorts in the United States, one associated with the state of Oregon, whose state animal is the beaver, and whose nickname is "the beaver state" (I wonder if they are also considering a name change?). Their state flag has a beaver on it—sort of. Here's the official explanation from the Welcome to Oregon (n.d.) website:

State Flag - The state flag is navy blue, with gold lettering and symbols. Adopted in 1925, the flag has the legend, "THE STATE OF OREGON" on the face, written above a shield surrounded by 33 stars. The 33 stars symbolize the fact that Oregon was admitted as the 33rd state. The shield, which is part of the state seal, has the number 1859 written below it. 1859 is the year that Oregon was admitted as a state. The reverse side of Oregon's flag shows a beaver. Oregon is the only state in the union that has a different pattern on the reverse side of their flag.

So, rather than go on about beavers and backsides, lest you think this discussion is getting badly derailed and otherwise, well, *flagging*, let's just say that the maple leaf and the beaver are both worthy symbols of national pride (I wonder what Canadian-born general semanticist S. I. Hayakawa thought about the matter?). As our not entirely forgotten friend Colby Cosh (2010) puts it, Canada has a dialectic between the beaver and the maple leaf (and no, there is nothing untoward about such interspecies relationships):

It would be a great shame indeed if one pole of this dialectic were broken and lost just because of the beaver$_2$ metaphor. I almost feel that *The Beaver*, as in the magazine, has been derelict in giving up the fight. But it's not my money. The brand could perhaps have stood up to any amount of silent snickering, but no media organ can afford to offer confusion to search engines and spam filters now. Google is a powerful, underestimated force for prosaicness: just ask any sub-editor who's been ordered to re-do a charmingly cryptic headline and get rid of the cute irony.

So, better to blame the internet, and Google, than to blame Canada, eh? But perhaps the solution lies in some combination of general semantics and the Semantic Web?

As Google gets smarter, or when some more subtle Nietzschean über-engine displaces it, such problems should disappear. Natural-language computing will, sooner or later, be able to tell whether a searcher or questioner is concerned with beaver$_1$ or beaver$_2$. The funny thing is, the beaver$_2$ metaphor may have long since disappeared by then, and the story of why *The Beaver* had to change its name may be an incomprehensible piece of trivia as far as the future is concerned. "Beaver$_2$" is already an outmoded vulgarism, something I would expect to hear spilling out of the mouth of a person my father's age rather than a 30-year-old. (Cosh, 2010)

An excellent point. Languages change over time, the meanings of words evolve, some words become archaic, and slang changes especially quickly, being essentially oral rather than fixed in writing. Egad! Zounds! Don't be such a ninnyhammer, blaggard, and winnit! Just pay attention to Mr. Cosh, please:

And we all know why: beavers$_2$ aren't seen in photographs or on video with their defining pelt very often anymore. Feminine depilation has become the norm—certainly not in everyday life, but in the visual culture. As a consequence, in all-male environments of the sort signified by the "locker room", the favourite metaphors of today emphasize different sensory aspects of the part in question. This is true even where the context involves a strictly visual encounter: the beaver$_2$ is not only called that because of its fur, but also because a gentleman is invariably excited to have caught a glimpse of one, as it were, in the wild. (Think about Jim Bouton's discussion in *Ball Four* of the baseball player's "beaver$_2$ shooting" hobby.) (Cosh, 2010)

And speaking of language, come on, we really have to hand it to Cosh for the subtle way in which he is handling the subject, successfully discussing the topic with tact and taste. And he continues

While reports of the extinction of the beaver$_2$ are much exaggerated, North America's experience of sudden shifts toward leg and armpit depilation suggest that women will change their grooming habits to match the visual culture. And not in the "long run", either, but in a matter of a few years. There is no evidence that there will ever be any mass anti-depilation backlash. If you are a traditionalist sort who has been waiting around for one, you have already been waiting quite a while. The arc of history bends toward bareness. (Indeed, it does so even when it comes to men's bodies.) (Cosh, 2010)

This is an excellent insight about the effect of what he calls *visual culture* on personal grooming. I would prefer to use the phrase *image culture* instead, because the contemporary media environment is not so much characterized by the ascendancy of vision, which after all is very much associated with reading, writing, and printing, as it is with the rise of visual *images* via mechanical reproduction, photography, motion pictures, video, and now digital media. In the field of media ecology, Marshall McLuhan and Walter Ong associate literacy and typography with visualism, while Neil Postman, Daniel Boorstin, Susan Sontag, Walter Benjamin and others concern themselves with the twentieth century's proliferation of images (see Strate, 2006).

And the depilatory phenomenon that Colby refers to strikes me as an extension of the value we place on cleanliness in American society, a value some identify as a myth of American culture, and certainly can be characterized as an overemphasis on the antiseptic, antibacterial, and antibiotic, and the deodorized as well (see Strate, 1982, 1986).

Anyway, Cosh (2010) concludes with a question:

> So it seems near-certain that we will, one day, be able to speak of the beaver$_1$ without the suppressed giggles. But will it still be associated with Canada, or will the symbol have been irretrievably relegated to the semiotic landfill?

I suspect that Colby is right, and I wouldn't be surprised if the beaver$_1$ symbol is retrieved and reclaimed by Canadians at some point in the future, and maybe even by Oregonians and Arcadians. But for now, these changes seem to suggest that we North Americans have the collective maturity of the *Beavis and Butthead* cartoon characters, so, what is left for me to say but, *DAM!*

Chapter Twelve

The Supreme Identification of Corporations and Persons

The Supreme Identification of Corporations and Persons

A news item that has generated a great deal of internet and media buzz is the January 21, 2010 Supreme Court ruling in *Citizens United v. Federal Election Commission* (130 S. Ct. 876) to allow unlimited corporate funding of federal campaigns. My friend and colleague Paul Levinson, who is a vigorous defender of the First Amendment, indeed, a First Amendment literalist cut from the same cloth as Supreme Court Justice William O. Douglas, has argued that the ruling was correct and beneficial (Levinson, 2010). But, while I don't share Paul's position on the First Amendment, that particular issue does not concern me here. Nor do I wish to go into the corrupting potential of unrestricted corporate spending on political campaigns, as did former Justice Sandra Day O'Connor, when she spoke about the potential for undue influence on the judiciary (on the *judiciary* mind you, beyond officials elected to the legislative and executive branches of federal, state, and local governments, see Mosk, 2010). Now, don't get me wrong, these are all very important matters, and all have been the subject of serious discussion, and ultimately these are the issues that matter in this case.

And yet, it seems that the lion's share of the attention, at least as far as I can ascertain, has centered around the fact that the ruling involved the legal decision to grant corporations the legal status of persons. This was a nineteenth century decision, traced back to the Supreme Court case, *Santa Clara County v. Southern Pacific Railroad Company* 118 U.S. 394 (1886). According to the rat haus reality press website, which provides a page devoted to the case (rat haus, n.d.), it was the result of a clerical error or unauthorized addition. Here's what they provide by way of preamble to the actual text of the decision:

> This is the text of the 1886 Supreme Court decision granting corporations the same rights as living persons under the Fourteenth Amendment to the Constitution. Quoting from David Korten's [1999] *The Post-Corporate World, Life After Capitalism* (pp.185-6):
>
> In 1886, . . . in the case of *Santa Clara County v. Southern Pacific Railroad Company*, the U.S. Supreme Court decided that a private corporation is a person and entitled to the legal rights and protections the Constitution affords to any person.

Because the Constitution makes no mention of corporations, it is a fairly clear case of the Court's taking it upon itself to rewrite the Constitution.

Far more remarkable, however, is that the doctrine of corporate personhood, which subsequently became a cornerstone of corporate law, was introduced into this 1886 decision without argument. According to the official case record, Supreme Court Justice Morrison Remick Waite simply pronounced before the beginning of argument in the case of *Santa Clara County v. Southern Pacific Railroad Company* that

> The court does not wish to hear argument on the question whether the provision in the Fourteenth Amendment to the Constitution, which forbids a State to deny to any person within its jurisdiction the equal protection of the laws, applies to these corporations. We are all of opinion that it does.

The court reporter duly entered into the summary record of the Court's findings that

> The defendant Corporations are persons within the intent of the clause in section 1 of the Fourteen Amendment to the Constitution of the United States, which forbids a State to deny to any person within its jurisdiction the equal protection of the laws.

Thus it was that a two-sentence assertion by a single judge elevated corporations to the status of persons under the law, prepared the way for the rise of global corporate rule, and thereby changed the course of history.

The doctrine of corporate personhood creates an interesting legal contradiction. The corporation is owned by its shareholders and is therefore their property. If it is also a legal person, then it is a person owned by others and thus exists in a condition of slavery – a status explicitly forbidden by the Thirteenth Amendment to the Constitution. So is a corporation a person illegally held in servitude by its shareholders? Or is it a person who enjoys the rights of personhood that take precedence over the presumed ownership rights of its shareholders? So far as I have been able to determine, this contradiction has not been directly addressed by the courts. (rat haus, n.d.)

And yes, this equation offers up all sorts of confusions, contradictions, and uncertainties. An article on the controversy in the online magazine *Slate* bears the title, "The Pinocchio Project: Watching as the Supreme Court Turns a Corporation into a Real Live Boy" (Lithwick, 2010). *The Huffington Post*, another online periodical, ran a humor piece entitled "Supreme Court Ruling Spurs Corporation To Run for Congress: First Test of 'Corporate Personhood' In Politics" (Klein, 2010). And the ruling was immediately subject to mockery on the part of the most trusted man in America (since Walter Cronkite, at least in certain circles), Jon Stewart, on *The Daily Show*.

Now, just to be clear about all this, the legal system does distinguish between a "natural person" (which is what the rest of us mean by "person") and "legal person," sometimes referred to as an "artificial person" as well. A "legal person" is a "legal fiction" which in a sense means that it's a fictional person, I guess… So, person$_1$ is not person$_2$ after all, to employ the extensional device of indexing recommend by general semantics, as a way to help us avoid the problem of identification that our language encourages. In other words, a legal person is not a natural person, and it's all legal terminology that may confuse the rest of us, but not jurists, right?

Well, that's in question here, since the decision seems to be about extending to corporations the legal rights of natural persons. As explained on the website called *The Straight Dope*, in response to the question, "How can a corporation be legally considered a person"

What most people don't know is that after the above-mentioned 1886 decision, artificial persons were held to have exactly the same legal rights as we natural folk. (Not to mention the clear advantages corporations enjoy: they can be in several places at once, for instance, and at least in theory they're immortal.) Up until the New Deal, many laws regulating corporations were struck down under the "equal protection" clause of the 14th Amendment--in fact, that clause was invoked far more often on behalf of corporations than former slaves. Although the doctrine of personhood has been weakened since, even now lawyers argue that an attempt to sue a corporation for lying is an unconstitutional infringement on its First Amendment right to free speech. (Adams, 2003)

Unless lawyers and judges have some general semantics education, officers of the court are just as prone as anyone else to make erroneous identifications. Call a corporation a *person* and you start to think of it as a person, and even if you started out with the distinction between the legal and the natural. Moreover, apart from the legal fiction, the very word *corporation* constitutes a metaphor of embodiment, the root *corp* meaning body, incorporation meaning to give the entity a body, to make corpo*real*. Of course, ordinary people do not, ordinarily, think of corporations as persons. We think of them as businesses, organizations, and several other things I cannot name here. It's folks working in the legal sector that talk about corporations as persons, and they are the ones who are vulnerable to falling into the trap of identification, of thinking

Corporation = Person,

when they should instead be thinking

Corporation ≠ Person.

Legal logic, like symbolic logic, allows us to make equations that make perfect sense in a given symbolic realm, universe of discourse, or deductive system, but that do not match up to the reality that we know through our senses, and our common sense. The legal map does not correspond to the territory, and the response on the part of critics has been ridicule, and rightly so.

And so, we find ourselves asking questions, some serious, some laughable:

If a corporation is a person, can it be subject to criminal charges? Actually it can, at least under some legal systems, although typically it's those natural persons who are corporate officers who are charged with crimes. Corporations can be fined, and sued of course, but can they be imprisoned? Can they be subject to capital punishment? Perhaps by being dissolved? Can they be arrested?

The Supreme Identification of Corporations and Persons

Can corporations vote? Can they run for office? Can they be drafted? Are corporations citizens of a nation? (They are, in a sense, citizens of the nation and state in which a corporation is incorporated.) Can they immigrate and become citizens of another country?

Can corporations marry? Jon Stewart suggested that the answer is yes, in the sense that they can merge. So corporations can marry each other, but can they marry a natural person? Could I marry Time-Warner, or Apple? Could we have children together? Well, maybe not in the biological sense, but maybe we could adopt? Can a corporation be a parent, or legal guardian of a child? (Isn't that what happened in the movie, *The Truman Show*?)

As persons, can corporations be owned? Clearly they are, but does that make them slaves? Can they be freed? (Many would say that corporations are our masters, not our slaves.) We might well remember that there was a time when African-American slaves were not considered persons in this country, or at best 3/5 of a person. Sadly, we have not always granted the status of person to all human beings.

If corporations are artificial persons, does that mean that robots can also attain that status? What is a corporation, after all, but a kind of intelligent machine whose parts include natural persons? (See Lewis Mumford, 1967, 1970, for more on this media ecological way of looking at machines; also see Peter Drucker's groundbreaking work, *The Concept of the Corporation*, 1946.) It's science fiction, sure, but this line of legal thinking opens the question up to serious consideration, after all.

If corporations are, in a sense, fictional persons, can other fictional persons also have rights? Like maybe Superman, Spider-Man, Mickey Mouse, and the like—except they are corporate trademarks anyway. Maybe Santa Claus, or Johnny Appleseed, or Tom Sawyer, or Paul Bunyan?

No, no, this will not do, not at all. I am reminded of a line from Kurt Vonnegut's novel, *God Bless You, Mr. Rosewater* (1965). In describing a fictional novel by the fictional novelist Kilgore Trout of a fictional future where suicide parlors have been introduced as a solution to overpopulation, he writes

> One of the characters asked a death stewardess if he would go to heaven, and she told him that of course he would. He asked if he would see God, and she said: 'Certainly, honey.' And he said,

'I sure hope so, I want to ask something I never was able to find out down here.' 'What's that?' she said, strapping him in. 'What in hell are people for?' (p. 22)

What in hell are people for? It's a question we don't have to reserve for the afterlife, or for God, it's a question we can ask ourselves, right here and now. One of our central media ecology scholars, Walter Ong, described his philosophy as personalism, the philosophy of the *human person* (see, for example, Ong, 1962). And it certainly makes sense to me to say, forget about these categories of natural and legal persons, let's just talk about human persons. And corporations are nothing more than extensions of human persons, meaning that they are inventions, technologies, media.

Of course, the whole reason why the doctrine of corporate personhood was adopted in the first place was to allow corporations to enter into contracts, to hire and fire employees, to buy and sell property, to be sued (which limits the liability of the stock holders and employees), and later to be subject to taxation. So, taking all that into consideration, if corporations are not called persons, what else might they be called? What other entities, aside from persons, can enter into legal and economic transactions?

The answer seems clear enough to me: Sovereign entities. States. Nations. Governments. Just as corporations are *incorporated*, that is, made into bodies, so too are states *constituted*, through their constitutions (a property they share with other organizations). The parallels are quite clear, but also quite discomforting. We quite rightly can be concerned, and fearful at the thought of the corporation as a sovereignty.

But corporations already are states, in effect. And rather than deluding ourselves with legal fictions of corporate personhood, let's work with the legal (and regal) realities. Let's talk about and think about and act upon corporations as nations. We will be, sooner or later. We can make treaties with corporations, and we can go to war with them if need be. This would make for a map that would better correspond to the territory, and it may well mean that we'll need new kinds of maps as well, to go with a new way of looking at the world that we already are living in. Korzybski (1950, 1993), who was a severe critic of capitalism and commercialism would undoubtedly approve.

Chapter Thirteen
Healthy Media Choices

Mary **Rothschild**: This Healthy Media Choices was recorded at the studio of WFUV on the beautiful Rose Hill campus of Fordham University, in the Bronx. Many thanks to WFUV for their hospitality. I have to say, it's easier to get here than to get across town in Manhattan, so another stereotype has bitten the dust. • My guest is Professor Lance Strate, Professor of Communication and Media Studies here at Fordham. He is the Executive Director at the Institute of General Semantics, and one of the founders and Past President of the Media Ecology Association. He is Past President of the New York State Communication Association. As a parent of an autistic child, he is involved in autism advocacy, and serves as an advisor for MOSAIC, a northern New Jersey support group for parents of autistic children. He is also on the Board of Trustees for Congregation Adas Emuno. He blogs (and I can tell you that his blogs are informative and enjoyable) about media, technology and communications. He's a poet, and has a poetry blog. He recently began podcasting. Last, but not least, he is one of the partners that make up NeoPoiesis Press.

Lance, you bring with you this incredibly rich understanding from your academic and your personal experience. You are the father of two children, your son is seventeen, your daughter is younger. I come from a totally different angle to the questions we're going to look at today. I come from early childhood education and parent education. I worked for years with young children. It was from that experience that I started doing parent education, because I was seeing on a daily basis what was being done to children's' attention. It has been a rich journey over the last ten years with Healthy Media Choices. And when you invited me to be on a panel co-sponsored by the Institute of General Semantics, I thought: "All these people are talking about the same underlying questions that parents and teachers have today." I want to focus on some of the aspects of general semantics, so, first, why don't you define general semantics for our listeners who may not be familiar with the field.

Lance Strate: General semantics is a system and a field and an approach that is concerned with how we relate to reality and how we understand

---

* **This chapter consists of an edited transcript of a radio interview conducted by Mary Rothschild.**

our environment, how we get information about the world, with particular interest in helping us to improve the accuracy of the maps we make of the territories that we encounter. That is one of the great metaphors in general semantics, the *map*, and the fundamental understanding that *the map is not the territory*. By this, we mean that the way that we describe things, the way we represent things with symbols, the way that we communicate about things through our languages or media, is distinct from our reality, it is not the same as what is out there, it is non-identical. And we need to remember that through our processes of perception and communication we only get part of the story, part of the picture of what's out there. Our maps can be more or less accurate, that is, they can be more or less similar in structure to the territory they describe, they can be more or less reliable in predicting what we will encounter, but they are always subject to error, and some maps may be completely erroneous. But there are ways that we can make better maps or better understandings of the world, become better mapmakers, and reduce the errors that our maps contain. General semantics is particularly concerned with improving our ability to evaluate our surroundings, and to pass those evaluations on to others as well.

Mary Rothschild: Would it be fair to say that a greater sensory palate with the environment is one of the ways that we get past a two-dimensional approach?

Lance Strate: Yes, it is only through our senses that we have anything approximating direct knowledge of our environment. In general semantics, one of the concerns is that, on the one hand we have this great gift of language, and our capacity for symbolic communication opens us up to everything that we consider distinctive about the human species, but on the other hand it also opens us up to a great extent to misconceptions and delusions. So we need to proceed much as scientists do when they talk about the empirical method, which means that they actually go out and look at things—they use their senses to check out the world and test their hypotheses. In the same way, we go around with our assumptions, our theories about the world. We need to go back to our senses and check them out, and then reexamine our assumptions and theories, and improve our hypotheses, and continue to do that, because the world is constantly changing. Nothing stays the same, we live in a dynamic reality. But we assign symbols to it, and the symbols don't change, the names stay the same, so we assume the world stays the same along with them. And that's not the case.

Mary Rothschild: That is such a crucial point, that we can so easily stop with a symbol, and think we know. This awareness that we don't know, the awareness that we've abstracted meaning into a symbol, and that it's not all contained there, is such an important thing that general semantics brings, and I feel that it is behind stereotyping each other because of skin, all kinds of things

Lance Strate: One of the great contributions that general semantics has made in the twentieth century has been education about stereotyping, helping people to understand the problems inherent in stereotyping, prejudice, and scapegoating, particularly in relation to race, religion, ethnicity, and gender. But stereotyping anything at all can be problematic, not just people.

Mary Rothschild: This is a burning question, for many people who've worked with young children for a number of years. There has been a radical change in the way children play. When young children are exposed to a lot of screen media, there's a deadening of imagination, a kind of stereotyping of play. There are references to screen media, completely out of context, in the middle of the day when they are trying to be attentive. This indicates a preoccupation with very strong images that come through media. So this has implications for something that is another concept in general semantics: "time-binding." What it really means, is this passing of culture through language and symbol through generations. So what do you see about that? What's your sense, talking about young children from birth through nine, the very youngest individuals, what's your sense of that stereotyping of interactions with the world in terms of culture being passed.

Lance Strate: Let me start with the first point. Children are already scientists of sorts. They naturally probe their environment, they test things out, maybe it's in a little thing like stepping on ants, or knocking over an anthill to see what happens. But when left to their own devices, they engage in a kind of play that involves gathering data and forming hypotheses. A lot of that productivity from play can be short-circuited through audiovisual media, through television, if they passively accept what they see on the screen. So I think that would certainly be a concern. From a general semantics point of view, we would want to encourage their active probing of their environment, and also direct them to improve their evaluations as they are going about doing this.

As to time-binding, that kind of opens up a fascinating door, because you go back to the roots of general semantics, and its founder Alfred Korzybski began by noting that what distinguishes the human species from other forms of life is our great capacity to preserve knowledge—that each generation doesn't begin all over again, but that we build up a storehouse of knowledge, and we're able to pass that on from one generation to the next, and in that way make progress over time. That storehouse of knowledge is dependent on our ability to communicate through language, through symbols, and especially through the very basic medium of writing, because that medium particularly allows for a tremendous buildup of knowledge. But even before writing, an equilibrium was achieved by building up knowledge through oral tradition. We find quite often in the modern world that tribal peoples are a wonderful source of new medicines, because what they do is probe their environment, they examine and evaluate the plants in their environment.

The great anthropologist Claude Lévi-Strauss pointed out that in oral cultures the individuals will have a name for every plant species in their environment. What they won't have are the more abstract terms like "bush" or "tree," that sort of vocabulary is missing, but they know every specific bit of flora and fauna in their surroundings. That knowledge is stored, is gathered, through a very painful and slow process, through many generations, and that is maintained very carefully by passing it down from person to person in a group-memorized, group-commemorated way. And when that process, that tradition gets disrupted, that's the end. There is no way to recover that kind of knowledge. What we as literates have done is to go to these people, and in a sense, mine their traditional knowledge in the way that we mine oil fields (to use a bad metaphor) and generalize from what we find, so we can take what is valuable, evaluate it, and also figure out what is not valuable—such as other parts of an oral tradition that might just be superstition—so we can separate those two out and benefit by collecting knowledge from all of the peoples of the world, to bring that all together.

There is an important place then, for oral tradition, and in fact all of our learning comes through in that way, from person to person speaking to one another across the generations. The general semantics position is that we can improve the ways in which we maintain our traditions, our cultural and intellectual heritage—it has never been in favor of eliminating our precious forms of time-binding. But you might say that the hidden and unconscious agenda of much of our contemporary media and technology has been to eradicate traditions by short-circuiting the process of communication across the generations.

Mary Rothschild: For me, this is essential. This very commercialized message, certainly in the past twenty-five years since the deregulation of advertising in children's programming, is coming in through children's programming, through even the most treasured programming stations. It is going to lead to some toy to buy. The average child spends over four hours a day with this. So just by replacing the time with people who might be telling them, modeling for them, another story, that commercialized story is short-circuiting the passage of culture. Is that your perception? Do you see that?

Lance Strate: I'd even hold commercialism aside, because that is a separate issue and an added concern. Even without commercialization, our media quite simply short circuit the process of traditional time-binding. That is, if there is a song or a game or a story that is passed on from one person to another, it's flexible. Each time we sing it or play it or tell it, it's at least slightly different. You may learn different lyrics, or a variation on the lyrics, the outcome of the game may change, the story may be condensed or elaborated upon. And part of what the child learns is that flexibility. But when you encounter the same thing in a mass-mediated form like a television program or a recording, it's always the same. It becomes fixed in a way that it is no longer part of *our* tradition, a tradition that we can use as we see fit or change or modify. It becomes *their* property, and it becomes fixed, and that becomes *our* problem. Then, on top of all that, bring in a commercial imperative, and that adds a whole other set

213

of problems, as the story or the song is tied to products that are being sold, there are ulterior motives, and that also alters the content. But still, I would go back to the primary problem, which is the short-circuiting of person-to-person communication.

Mary Rothschild: That is so important on so many levels, Louis Cozolino's book *Neuroscience of Human Relationships* (2006), points out that that there is this social-synapse that happens between us, as people. We are actually forming our brains as we are speaking and as our impressions of each other happen, and that doesn't happen with a screen. And there's the work of Patricia Kuhl and others at the University of Washington, where a child has a caregiver who speaks another language, and the child will pick up the cadences of that language and begin to mimic and understand, but, if a similarly aged child sees it on a screen, they don't pick it up (Gopnik, Meltzoff, & Kuhl, 2000). We think of screens as delivering something, but they also block, because that personal interaction isn't there. I think that is one of the things that even the most wonderful content—content is very educational, or very stimulating, or very beautiful—the medium of the screen is not the ideal one for a young child, or probably for any of us.

So where do we go from there? If we communicate in this uniquely human way, between generations, through language, symbol, and story and the story is being usurped and truncated, then how do we bring awareness to that? I think general semantics is in a great position to do that, but how do we raise awareness and not be alarmist? So many people think that if you wish to cast light on this, you are saying "It's evil, it's bad," this whole alarmist thing which manipulates parents as much as anything does, how to not do that, but say, "Let's look at what is actually needed and healthy here, and let's look at what is actually going on, and make some informed judgments." There is this sense that we're like *Prevention Magazine* thirty years ago, and now you can't find Wonder Bread, that we're trying something very difficult against the mainstream. But how to bring these questions in a way that isn't a "he said, she said," "yes vs. no" conflict, which from my point of view is not what's needed. We need to come in an impartial way, which is what I value about general semantics, it brings this kind of wonderful impartiality, and says, "Let's look at it." Do you see anything emerging around this?

Lance Strate: I think one area that is hopeful has to do with what we call new media, and digital media, in the sense that they open up more possibilities for user control, at least teaching about the fact that there are camera angles that are chosen, that the point of view that you are getting is not necessarily *the* point of view. I think we can do a lot more to bring out the understanding of how things are produced. It helps to be able to produce things by yourself. That is one side of media literacy—to learn how to control the production side of the process. It is also what's been hailed as the wonderful thing about the new media—the fact that you can create video fairly easily, you can create your own podcast, you can create all the things that once we could only consume, and we can create what we ourselves want to consume, rather than be limited to what commercial interests are willing to provide for us. Clearly this is not enough in and of itself, but this addition is certainly one step in the right direction. And while it's true that technologies open up these new possibilities, there is still an ongoing battle between people who want to skew the technologies in a consumption orientation, and those who want to open them up so that everyone can take part in production.

To answer your question in another way, there was a well known Brazilian educator, Paolo Freire, who wrote a book that was very popular, and insightful, *Pedagogy of the Oppressed* (1970). Part of his argument is that education cannot be truly liberating if it's done in a hierarchical top-down manner. Rather, it has to be approached in an egalitarian manner, with students and teachers participating as equals. So, Freire would go around and give lectures, and people would often ask "Aren't you violating exactly what you are advocating in lecturing to us?" And in a way, that is part of the question. Can you raise awareness, can you teach people about what we've been talking about, other than by doing so person-to-person? It may not be possible; it may require that we do it face-to-face, voice-to-voice, that may be what's needed. Once again, just to look for signs of hope, one of the great things about the internet, in its early iterations, with email for example, and more recently with social media, is that it's opened up channels of interpersonal communication. It's not quite as good as face-to-face communication, but it has allowed people to connect who were otherwise unable to connect. That at least creates some possibilities for awareness and for education. We're not going to get it by having promotional spots; we're going to get it by having people connect on a one-to-one or small group basis.

Mary Rothschild: I certainly can attest to the fact that Twitter, and Facebook are great gathering places for teachers and parents who have these questions. There are lots of conversations happening there about what's going on: how to have our children be freer (why are we so afraid when statistics actually say that children are safer now than they used to be?) And getting back to your point about using the media, one of the things we advocate at *Healthy Media Choices* is telling the family story through media. Having the child see media tools being used intentionally can counterweigh a lot of other things because then they have a firsthand experience of how it's done, how it's manipulated, how it's edited, but also they are seeing the story evolve, they are maybe participating in telling it. There's a whole delicious mix.

Lance Strate: And they are seeing what's being left out. They are seeing the editing process, they are seeing the selection process, and that is a key element in general semantics, to understand how we *abstract*, how we select, and leave out different elements as we create our messages. That is so vital.

Mary Rothschild: It would make a big difference if people who did children's programming were required to do some family media literacy. Who made this? How is it made? This is basic media literacy. Why did they make it? What was left out? How were the camera angles chosen, what would it have been like if it had been slightly different, or only one camera? All of those things can really be part of the family conversation in an informal way, it doesn't have to be negative, it could be interesting.

Lance Strate: And actually, just recently, I got approved for a grant by the Time Warner Cable Research Program on Digital Communications, together with my colleague Lewis Freeman, and we want to create a plan for how children's television in the future could incorporate media literacy education, how it could use digital technologies to allow for exactly what we have been talking about here. That is, instead of the child just passively watching a television program, make it possible for him or her to rewind and replay in different ways, choose camera angles, get information about the production, etc., and that would make children's programming not just educational, but encourage media education, and media literacy. I'm very quick to say that is only part of what can be done to improve our current situation, because the other part is to bolster family communication, direct communication. The problem is that it's not just a matter of saying, "You should talk to your kids more!" That is somewhat obvious, but the problem is to a large extent situational, it is subject to the economy, to social issues and interpersonal issues like divorce. There are so many factors that feed into this, and we really have to look at the big picture, at the ecology of all these different factors. You can't just have media literacy in isolation to everything else that is going on.

Mary Rothschild: That is why in our workshops we talk about the household as an ecosystem. It isn't just family. There are children who are there part time, there are caregivers who come and go. Who is in that ecosystem? What are their needs? If you have an elderly person living with you, who feels they have to have the television on all the time, that is a very different situation from a couple with a new baby who has nobody else living with them; there's carte blanche to control the environment somewhat. So we're really looking at who you have in the household and what the needs are. If you have a teenager, there are some very real needs that can be just as serious as those of a young child. So to balance all of that, as many of those people as possible need to come into the conversation about how to spend time together, what the priorities are, and then make some informed choices and be intentional. I think the thing that so many of us are seeing is, with the demands of the economy, people are working a couple of jobs, kids are in so many activities very often. There is a lot of stress, and the kids are stressed about school, tremendous stress even on very young children. So then leaving time, leaving space, finding where there already is time spent together is so important.

There's a wonderful story from Keith Frome who wrote a book *What Not to Expect: A Meditation on the Spirituality of Parenting* (2005). And he tells this story about how he used to drive his son to school. He was the headmaster of the school. His son was late, not coming downstairs, one morning. Keith had a meeting, so the caregiver for his younger child said, "Why don't you leave him, you'll teach him a lesson, he can walk to school." So he left, and it was only when he was in the car that this feeling of mourning came over him, that he realized that this was the one time when he and his son were consistently alone during the day, and it was a ritual that had value that he hadn't seen before. So, it's more or less mining our days for that time, and making our choices about turning everything off, or being more intentional about actually communicating with each other. It's not about reinventing the wheel, so what you're saying is so important. All of these things: culture, information, media literacy education, are, essentially, one on one, most effectively, and then, as it grows, it can become a larger societal conversation.

I noticed that you are on the board of trustees of your synagogue, and I am interviewing people from different faith and community organizations, about what communities are seeing about children, and the culture of the communities, because so often stories of culture are in our faith communities, in our humanist communities, and they have an imperative to pass that culture on to the children. What I'm hearing is, not so much from the people I'm interviewing from the communities, but from educators, for instance, in Jewish private schools, is that even though there is this tremendous culture of care—the families are there, the values are there—there is all of this commercialization coming in and it is such a strong dynamic on the children, that it is very difficult to get them to be attentive to stories of their culture. What I'm seeing in my interviews is that this concern about media, children and culture doesn't seem to be filtering into the faith and humanist communities as an imperative. What do you see about that?

Lance Strate: It is extraordinarily different these days. When I was growing up it was so very different, because there was this strong sense of "you have to do this, you have no choice!" And that created a tremendous backlash, it certainly could be alienating, but you definitely knew where you stood, and you knew this was important to your parents, to your family and your community. I grew up in the tradition of Reform Judaism, and that sort of thing was in place even in that relatively liberal movement; but now, in the Reform congregation that I am involved with, you find people almost apologizing for asking young people to take part in the tradition, saying, "maybe you'll want to do this in the future." There isn't this sense of "this is important," because no one quite knows what is important anymore, no one is quite sure of how they feel about things. Obviously in more orthodox traditions that is not the case, and then it is quite easy to enforce the separation, and even hostility and rejection that you see in many of the fundamentalist faith communities towards the mainstream media. For those of us who are on the progressive side of things, it's a very difficult proposition. Actually a colleague of mine up in New Paltz, Donna Flayhan, makes a wonderful point, that what we all need is a Sabbath, in particular a Sabbath from media. She points to the Jewish Sabbath, her background is actually Christian Lebanese, and the Jewish tradition is to take a break from electricity and from work, which in turn forces a break from much of our media, and we need that. We need to say, "Let's not watch television for a day, and let's not go on the computer, let's not check our email and Facebook." And I do think that is something we need to think about; we need to reconnect with that aspect of our religious tradition. And there's that problem again, going back to time-binding, that if you follow the strict path and say we have to reproduce the past as faithfully and traditionally as possible, then you don't make any progress, and you don't adjust for changing circumstances. The great challenge is how to maintain a tradition while adjusting for changing circumstances, and it is tough because what is offered by media like television and the internet in the place of our traditions is so seductive. We are offered such wonderful entertainment and distractions from the more arduous processes of intellectual development and spiritual growth. But I think most of us know, most of us sense the emptiness within our media offerings, not that they are evil, but that they're not enough for us, that we need something more, and we are not sure how to get it, how to find it. And that is where a progressive faith tradition needs to step in and say, "*this*, *this* is what you are missing, *this* is what is not there, *this* is where you find meaning, and a connection to something greater than ourselves, something greater than just working and buying stuff and playing around. *This* is the place to look for it."

Mary Rothschild: For so long the very conservative religious groups have owned the issue of media's impact on children, or are perceived to own it. It is one of the difficulties of speaking with people now. There is a hesitation, a fear of being identified with that. For those of us who are trying to find a third way into the question without setting up the backlash that you are talking about, that's a difficulty. With *Witness for Childhood* we are clear—these are progressive voices. It is not about passing on orthodoxy, in fact I would say one of the basic things is just passing on the ability to be silent, the ability to be still at all and come to ourselves. (We use a metaphor about attention being a muscle we use to come into ourselves.) All religious traditions use this ability.

Lance Strate: Certainly, one of the significant aspects of religious traditions involves the structuring of consciousness, and developing consciousness. Absolutely—meditation, prayer, these are exercises in expanding consciousness. Working at Fordham University, I've become familiar with the Jesuits' tradition in Catholicism, and the founder of their order, Ignatius Loyola, introduced a set of spiritual exercises. And I think this sort of thing extends to all faith traditions, facilitating a form of personal growth; in the negative sense, we could talk about cults that manipulate and brainwash, and all that, but in the positive sense it is about raising consciousness, bringing about a higher state of awareness and equanimity.

Mary Rothschild: That's the essential aspect. And when you look at a child's day, x number of hours at school, and four hours in front of the television or computer, where is that child ever still? The Sabbath would be a huge step. There are a lot of people who aren't associated with any community, but who could still share that with their child, a time, a Sabbath - perhaps another word—a time when everything extraneous is shut down. Perhaps the time is spent in nature, we do an exercise in our workshops where we try to remember a time as a child when we *knew we were there*, and nine times out of ten it is an experience in nature.

Lance Strate: I think that is very important, especially today with so much grave concern over the environment. We need to include the natural world as the medium for spirituality. Quiet, and patience is part of it, silencing ourselves, externally and internally, so that we can truly listen to what our environment is telling us. And seeking, understanding that we are seekers. On the progressive side, we don't claim to have absolute truth. There are great mysteries out there, and we are all seekers on a path towards greater enlightenment. I think it goes back to the idea of probing our environment. We do that in a scientific way, but we also do that in spiritual way—through meditation, through prayer, through communion, or spiritual activity. We are probing, we are seeking, we are trying to understand. And that is what it is all about.

Mary Rothschild: You know what comes to mind from some of the things you said is Bill McKibben's work *The Age of Missing Information* (2006). It's not just information from the world around us that we are missing, but information about ourselves. And so much we can learn about ourselves is from communion with nature, and being still. But I can tell you that children, even as young as three years old, can be in nature for a long period of time, and suddenly they'll say something like "The Lion King Video is too loud." And you can say, "Where are you hearing the Lion King Video?" And they say, "In my head." The sensory information about themselves and the environment is intercepted by these strong images. Mary Catherine Bateson mentioned something like Sabbath when I interviewed her recently, so I hope this idea is starting to get some tread.

Lance Strate: Well it's kind of like when the toddler learns the word "no" and then goes crazy for a while, going around saying, "no, no, no." In a sense. that is when we become fully human, by which I mean that "no" is the most abstract of terms. There is no way to depict "no", no picture we can use to represent nothing or negation, no way of acting it out. It only exists as an abstract concept. Whether it is in the word "no" or the circle with the line across it, or the nonverbal symbol of shaking your head, you can't show it, can't demonstrate it. So just saying "no" is an essential part of what makes us human. I don't want to go to the extreme of the strict "no!", but we do need the idea and practice of self-discipline, knowing that we are capable of turning things off and saying "not now." That really is the key to much of what we're talking about here.

Mary Rothschild: And it's not just the child we're talking about here. The research from the people at the (now defunct) Center for Screen-Time Awareness, the TV turnoff people, showed that at the end of the TV free week, it was the parents who turned it back on. The children were having a great time. They want the attention of the parents. I totally understand that, there are so many demands on parents. In the back of your head, you're thinking about the bills you have to pay and the chores that need doing.

On the Binding Biases of Time

Lance Strate: Sure! There is so much stress in our lives. I said we need a lot of help, and I don't mean that in a mean way, things are very hard for people today, at least in some ways. I know the "Greatest Generation," they went through the Depression, and then went off to the war and were very heroic in facing up to that, but they also had tremendous support, community support, family support, and that just doesn't exist in the social structure today. Everyone is off on their own, and we don't have the resources to deal with a lot of things coming our way, and that turns us towards television and towards the internet instead. There are great benefits from having these sources of information available to us, but we're missing something, it goes back to the fact that we're missing something.

Mary Rothschild: We're missing, from my point of view, the ability to look at it. Taking the time to look at the situation, see what you want, and brainstorm strategies with other people, which is what we try to offer. So many people say: "I plug her in because I have to make dinner." Well, that child can be along side doing something age appropriate to help prepare the dinner.

Lance Strate: And the irony is we spend less time making dinner than past generations. And it used to be that making dinner was something children could be a part of, or at least be connected to in certain ways, but now they are disconnected from it, or alternately, just told to go heat something up in the microwave for themselves. But the last thing I want to do is blame the parents, because that's so easy. You always know the people without children because they are the ones blaming the parents, not the folks with children of their own.

Mary Rothschild: The difficult part is not to have parents feel blamed or feel guilty when they look at what's happening, because they might be horrified if they actually calculate the number of minutes they spend with their child. We're trying for a more impartial look—this is what is, this is what I want, what can change? Where can I affect something? There was one woman who stands out in my mind. We make these refrigerator magnets in our workshop that remind us about what we thought we were going to do. (I don't know about you, but I can go to a workshop and when I walk out the door all my habits meet me and it all goes out the window.) And this one woman, she lived in a very small apartment with a partner who was not going to turn everything off, and she said, "Just go for a walk once a week together." She felt that was something everybody would buy into, and I thought that was a really successful move for that woman, because she found something that she thought was doable for her ecosystem, something they could build on. They can build from those sensory experiences and conversations they would have alongside each other. So the last question I wanted to ask you is, if you were speaking to faith community, they came to you and said "With your understanding as a parent and an academic in this field, how could we approach these questions with our parents, and also, what role could we have in a larger society to bring this kind of question to light?" What would you say to that?

Lance Strate: I'd like to talk about my tradition if I may. Judaism is a tradition that we've been maintaining, that goes back about 4,000 years. And that's really something, it's really astounding when you think about being part of a continuum of four millennia. And I raise the question, isn't that something that is meaningful and something that we would like continue to be a part of, in and of itself? The vastness of it boggles the mind, and I think that kind of perspective, and that kind of understanding is worthwhile, because it's also presenting us with a historical context. What is going on today is not everything. It reminds us that we owe a tremendous debt to the past, that most of what we have today we did not earn, it's an inheritance. We didn't create it, we didn't make it, we would be helpless without it, if we had to come up with it from scratch. It was the hard work and the lives lost in the pursuit of knowledge and an improved way of life, it was the efforts of untold generations that brings us to the point that we're at today. So we have to appreciate that, and in appreciating that, we also *must* preserve it—it is an obligation, not an option—and pass it on to the next generation, and not lose it due to our own stupidity.

Mary Rothschild: This is a good place to end. In my view, in a family where the adults have that kind of appreciation and passion, the children will absorb that directly and in the future, the question the progressives have about it possibly not being forever, that will take care of itself. Something will be implanted in the child that values the essential actions of ritual and actions that connect us with our higher selves, and with each other. Thank you so much Lance, I hope we can speak again.

Lance Strate: I'd like that.

Chapter 14

The Future of
Consciousness

A brief introductory note: My topic, "the future of consciousness," was given to me by Allen Flagg,[*] and it is a challenging assignment, but I believe I have a good enough handle on the subject of consciousness to have a few things to say about its future. Much of what I have to say will be a review of the past and present of consciousness, however, in the hope of establishing a basis for considering its future, so I hope you will bear with me. And, if you don't like what I have to say, it is my sincere hope that you hold Allen Flagg personally responsible for whatever defects you happen to identify.

Consciousness is a curious topic to consider, for it at once places us in the realm of self-reflexiveness. It is the mind thinking about the mind, which is very much akin to the blind leading the blind. I am at once reminded of the biologist Lyall Watson's self-reflexive paradox that "if the brain were so simple we could understand it, we would be so simple that we couldn't." [**] It is a basic tenet of systems theory that you cannot completely understand a system from within. The whole is greater than the sum of its parts, and you have to step outside of the system to see that whole (see, for example Bertalanffy, 1969; Capra, 1996; Laszlo, 1972, 1996). I try to think of this whenever I am coming over the George Washington Bridge to New York from my home in New Jersey and I find myself stuck in what appears to be a completely random and unexplainable traffic jam. After all, you have to do something to pass the time when it's bumper to bumper. So I try to imagine a distant, bird's eye view of the system of roads and highways that surround my position, knowing that I would in all probability see a rather orderly traffic pattern, a rhythmic pulsing along the arteries of New York. Of course, if I could just step outside of the system, I wouldn't be stuck in traffic and forced to contemplate these things in the first place. The problem is that sometimes there is no way to get outside of the system, sometimes you find yourself on a highway with no exits, truly an existential dilemma.

And that is why it is impossible for us to imagine or truly understand

---

[*] As the topic for an address at a general semantics symposium.

[**] I originally came across this quote in *More on Oxymoron* by Patrick Hughes (1983, p. 145). The quote was only attributed to "a philosophically minded biologist," but a quick internet search revealed that Lyall Watson was the biologist's name, although I was unable to find a specific bibliographic citation for his quote.

death. We cannot step outside of our own existential system. So many of us cling to the notion of life after death because we cannot imagine death after life. We can't leave the system, so we imagine more of the same system, more or less. But to insist that there is no life after death would be equally unimaginative, as we can neither confirm nor deny what lies beyond our system. The only certainty is uncertainty.

The problem we face in asking the mind to reflect back upon itself in these various ways also relates to the mathematician Kurt Gödel's incompleteness theorem, which shows that no mathematical system is entirely free of contradiction or capable of proving its own validity—there must always be recourse to something outside of the system (see Hofstadter, 1979, for an insightful and entertaining discussion of the incompleteness theorem and recursion, aka self-reflexiveness). And this also brings us back to Bertrand Russell and Alfred North Whitehead's theory of logical types, and the premise that a category cannot be a member of itself, unless you want to invite paradox (Whitehead & Russell, 1927-1927).

So to avoid paradox when we think about consciousness, we need to move from inside the category to outside of it, to consciousness about consciousness. But how do we "go meta," as Douglas Rushkoff (2006) puts it, how do we get outside of ourselves? Now, I raise this familiar question as a ploy to gain your sympathy, and to impress upon you the difficulty of what I am attempting to do in this essay. General semantics asks us to maintain an extensional orientation, that is, an orientation based on observation of the outside world, not on our own internal logic, or feelings, or prejudices. But consciousness is an entirely internal, intensional phenomenon. Our understanding of consciousness is based on our subjective examination of our own consciousness, and on inferences and logical reasoning in lieu of empirical evidence. In other words, there is no way to avoid applying an intensional orientation to an intensional phenomenon.

Further complicating the matter is the fact that the term "consciousness" refers to a number of different concepts, so that we would be well advised to use the general semantics device of turning singular terms into plural ones, and speaking of consciousnesses. For example, one type of consciousness that is popular in the New Age movement is the idea that consciousness extends throughout the natural world (see, for example, Campbell & Moyers, 1988). Even a stone may have a consciousness of sorts, in terms of its own integrity as a stone. This idea about consciousness is an updating of paganism and animism, in which the whole of nature has a spiritual dimension, so that stones, and rivers, and trees, and the sun and moon have minds, and affect us by force of will and supernatural agency. This understanding of the world, by the way, should not be dismissed too lightly. It is a highly efficient form of theory-making, a theory of mind (Baron-Cohen, 1995; Frith, 1989). Thinking of lightning as having a mind of its own, as being an agent who is angry and threatening, is actually a pragmatic and effective way to relate to a natural phenomenon. It has survival value, which is why the anthropomorphic view of the world has served us well for most of the history of our species.

Related to animism are other notions of supernatural consciousnesses, spirits, nymphs, gods and other conceptions of the divine including those that appear in Judaism, Christianity, Islam, Buddhism, and rest of the organized religions. And there are also various ideas about human consciousness surviving death and continuing on in a disembodied state, on earth or in heaven, not to mention related concepts of psychic phenomena. Further, there are variations on the Gaia hypothesis in which the earth is seen as not only a living organism, but one possessed of its own spirit and consciousness. It is not my intent to ridicule or dismiss matters of the spirit, or to endorse them; I only mean to describe them as one approach to consciousness.

A parallel can be drawn between the universe of the spirit and the equally ethereal universe of information. Information is not a material thing, but rather as conceived of by Claude Shannon it is the quality of negative entropy, that is of order in the face of chaos (see Shannon & Weaver, 1949). Information systems process data with the effect of increasing the organization and differentiation within a system. This is a property of all living organisms. And it is also a property of advanced technological systems that employ feedback loops, as Norbert Wiener (1950, 1961) explained over a half century ago in his books on cybernetics. Since then, we have grown accustomed to the idea that certain types of computer programming are called artificial intelligence, and we have come to refer to certain technologies as smart—smart bombs, smart houses, smart clothing, etc. It is not surprising to learn, then, that within the relatively new field of cognitive science, cognition is seen as a function of any information system, be it human, computer, or insect. Even the immune system can be described as employing a form of cognition as it goes about its task of differentiating which cells do and do not belong. While cognition is not exactly the same thing as consciousness, the presence of one implies the possibility of the other. Stanley Kubrick's masterpiece, *2001: A Space Odyssey,* famously posed the question of whether artificial cognition constitutes artificial consciousness. And the same question can be directed towards the communal functioning of an ant colony, and the autonomic functioning of an organism.

In this sense, we can add to the ideas of consciousness as spirit and consciousness as information a third concept of consciousness as life. Life is at once the combination of spirit and flesh, and a process that is fundamentally negentropic, cybernetic, and cognitive. Consciousness can be understood in terms of life force, and in terms of the basic functions performed by all living organisms. In particular, consciousness is associated with the organism's observable responses to outside stimuli. If the organism reacts to stimuli, it is conscious; if not, even if it is still technically alive, it is thought to be without consciousness. Consciousness also tends to be associated with some type of purposeful activity, even if it is the simple movement towards light and away from darkness.

The more typical approach, however, is to limit consciousness to those living organisms that have nervous systems. It then becomes a living function of nervous systems in general, and specifically of the brain. In this way, consciousness becomes localized to one subsystem of the larger system we call an organism. Consciousness then can be operationalized and measured in terms of neural activity. Organisms with nervous systems require a minimal level of neural activity to keep their autonomic systems going and remain alive. But consciousness is especially associated with a high degree of neural activity, as well as responsiveness to outside stimuli. As neural activity decreases, and responsiveness is reduced or disappears, we say that the organism moves from a state of consciousness to one of unconsciousness.

This introduces an insoluble problem into our understanding of consciousness. If we base our concept of consciousness on responsiveness to outside stimulation, where do we draw the line between consciousness and unconsciousness? At what point do we say that responsiveness has been so reduced as to have crossed over from consciousness into unconsciousness? And at what point does unconsciousness cross over from something like dreaming to something more like death? Even if the organism is completely unresponsive, the presence of a nervous system suggests that consciousness could possibly persist internally. We look for answers by measuring neural activity, but measurement holds no answers about cut off points, quantification cannot help us answer qualitative questions. Sure we can operationalize and say that a given level of neural activity is the definition of consciousness, but it would be a mistake to reify or idealize a relatively arbitrary definition. In truth, it could be argued that whenever there is some level of neural activity, or even whenever there is a living being that responds in some way to stimuli, there is some form of consciousness. This is no small problem, as the recent controversy concerning Terry Schiavo brought home. As in the case of coma victims, the conceptions of consciousness based on neural activity, on information processing, and on spirit are difficult to separate from one another.

It does seem that consciousness is connected to biology, but at the same time not the same as biology. The metaphor that is sometimes used, and that I find resonant, is that of fire. Just as wood can serve as the material base of fire without actually being fire, the brain serves as the material base of consciousness but is not itself consciousness. And those whose minds have lost their fire are called deadwood. The fire metaphor is particularly apt when you consider how the brain is entirely dependent on oxygen, so that a brief interruption causes us to "lose consciousness," while a longer interruption snuffs out the flame of life. The fire of consciousness also seems much like the burning bush of Moses, which burns but is not consumed. Perhaps it was actual fire that ignited consciousness, for we often consider the taming of fire to be a defining characteristic of human evolution, the moment when we master technology and also the principle of transformation. According to Claude Lévi-Strauss (1969), fire not only represents the transformation of nature into culture, but as such represents our entry into symbolic communication.

Sigmund Freud (1977) separated consciousness into the conscious and unconscious minds. Unconsciousness in psychoanalysis becomes an alternate state of consciousness, of the sort we associate with dreaming or hypnosis. Put another way, Freud viewed the unconscious as actively engaged in information processing and cognition, as well as being a biological and neurological phenomenon. For Freud, consciousness was associated with the ego that emerges during the waking state, while the unconscious mind is always lurking in the background, and leaking into the foreground in the form of slips of the tongue, neuroses, and such. Viewed from another perspective, however, we can recognize that consciousness encompasses the unconscious as well as the conscious mind.

Sleepwalking occurs when the unconscious mind takes over the body, but sleepwalking is also a metaphor. Being asleep symbolizes a lack of awareness and attentiveness. And from along these lines, consciousness can be associated with our ability to actively think about phenomena, to actively pay attention to our surroundings, to actively respond to stimuli from our environment. It is a state of being alert and mindful.

It is a short step from awareness to ideology, and the Marxist notion of consciousness (see, for example, Eagleton, 1991). For traditional Marxists, consciousness translates to a political understanding of the world. The consciousness of the ruling class naturally rationalizes their good fortune, and views the subordinate classes as getting exactly what they deserve. This type of consciousness is called a false consciousness. And it is transmitted to the subordinate classes as ideology through institutions such as schools, churches, the arts, and the mass media. In this way, the subordinate classes adopt the false consciousness of the ruling classes. The answer to this problem is to teach them the truth, teach them about the reality of their oppression, and thereby raise their consciousness. The idea of consciousness raising was adopted by feminists as well to describe their efforts at ending gender discrimination. I should add that over the past half century neo-Marxists have abandoned the dichotomy between a false consciousness based on ideology and a true consciousness based on reality. They replaced it with the relativistic notion that all consciousnesses are a function of ideology. It is all false, in the sense that it's all a social construction, and all that matters is whose side you're on (see, for example, Williams, 1977).

Notions of false consciousness and raised consciousness, while broad in certain respects, appear shallow in contrast to Freud's depth psychology. Certainly, they are closer to the surface than notions of altered states of consciousness. As pioneered by Timothy Leary and his followers in the psychedelic sixties, hallucinogenic substances were ingested in order to alter sense perception and thought processes. From this perspective on consciousness, different states carry with them different modes of awareness, so that by experiencing a greater variety of states of consciousness, opening the doors of perception as it were, we can increase our overall level of awareness. Often this involves moving from a linguistic, intensional mode to a perceptual, extensional one. Again, my purpose here is not to endorse or condemn, only to describe.

I should note that to some extent religion has always been concerned with altered states of consciousness, and with generating higher levels of awareness. Mystical experiences fall into this category, but we can also see the activities of prayer and contemplation as attempts to mold the thought processes and consciousnesses of adherents. Art too can be a means of achieving new modes of sense perception and alternate states of consciousness, a point that Marshall McLuhan (2003; McLuhan & Parker, 1968) was fond of making. And Neil Postman (1979; Postman & Weingartner, 1969) would remind us that teaching is most certainly an effort to shape thought, perception, minds, and modes of consciousness. Education is a consciousness raising activity, as is general semantics itself, as we seek to raise awareness of language and symbol use, and instill consciousness of abstracting. Of course, general semanticists are also known to say, "get it into your nervous system!" which reflects an additional biological perspective.

Many of these ideas relate consciousness to uniquely human modes of thought, and to higher mental processes. And more than a matter of simple awareness, a distinctively human consciousness can be viewed as an awareness of one's own self, an awareness of one's own awareness. We are self-aware, self-conscious, that is we are conscious of our own consciousness.

Awareness of our own consciousness is intimately caught up with awareness of the consciousness of others, with theory of mind, the theory that others have a mind that works more or less like our own. A classic example of theory of mind goes like this. I tell you that two individuals, let's call them Stuart and Wendell are in a room together. Together they put a fifty dollar bill in a box and close it, and then they both leave the room. A little later, Stuart sneaks in on his own, removes the fifty, closes the box back up, and leaves. Then Wendell returns to the room. The question that is then asked is, where does Wendell think the fifty dollars is located? The answer we typically give is that he will think it is in the

233

box. We assume he has a mind just like our own, we empathize with him and understand what he can and cannot know based on his limited point of view, and we understand the act of deception committed by Stuart. Individuals who are blind and cannot visualize the scene still come up with the same answer. Individuals of limited mental capacity also understand the difference between what they know and what Wendell knows. It is only individuals who are autistic who will answer differently, and say that Wendell will know that Stuart has the money (Baron-Cohen, 1995; Frith, 1989).

According to Simon Baron-Cohen (1995), autistic individuals are characterized by mindblindness, his way of referring to the failure to develop theory of mind. They do not see others as having a mind like their own, and have relatively little or no self-consciousness. Does this mean that they do not have consciousness altogether? I have to admit to a certain lack of impartiality on this matter, as my daughter is autistic. From what I have observed, she thinks, she plans, she plays, she jokes, she gets angry, sad, and frightened, and she loves. She even constructs novel sentences, despite a very limited capacity for language. But her self-awareness and her understanding of others has not developed the way it has for typical children.

Donna Williams (1988), a high functioning autistic, describes her own view of the world as follows:

> Up to the age of four, I sensed according to pattern and shifts in pattern. My ability to interpret what I saw was impaired because I took each fragment in without understanding its meaning in the context of its surroundings. I'd see the nostril but lose the nose, see the nose but lose the face, see the fingernail but lose the finger. My ability to interpret what I heard was equally impaired. I heard the intonation but lost the meaning of the words, got a few of the words but lost the sentences. I couldn't consistently process the meaning of my own body messages if I was focusing in on something with my eyes or ears. I didn't know myself in relation to other people because when I focused on processing information about "other," I lost "self," and when I focused on "self," I lost other. I could either express something in action or make some meaning of some of the information coming in but

not both at once. So crossing the room to do something meant I'd probably lose the experience of walking even though my body did it. Speaking, I'd lose the meaning of my own sounds whilst moving. The deaf-blind may have lost their senses; I had my senses but lost the sense. I was meaning deaf, meaning blind. (p. 33)

What Williams describes can be considered an alternate state of consciousness, one that typical individuals might achieve by artificial means. Altered states of consciousness achieved by meditation or drug use often involve the retrieval of a prelinguistic state of mind, and in this respect, it is instructive to consider how Temple Grandin describes her condition. Grandin is a professor of animal science at Colorado State University, a designer of livestock facilities, and probably the best known autistic individual in the world today. In her own words,

I think in pictures. Words are like a second language to me. I translate both spoken and written words into full-color movies, complete with sound, which run like a VCR tape in my head. When somebody speaks to me, his words are instantly translated into pictures. Language-based thinkers often find this phenomenon difficult to understand, but in my job as an equipment designer for the livestock industry, visual thinking is a tremendous advantage. (Grandin, 1995, p. 19)

When it comes to intelligences, autistics defy normal categorization, as they may be savants in one or two isolated areas, and far below average in others. Based in part on this phenomenon, Howard Gardner (1983, 1993) arrived at his theory of multiple intelligences, and I would suggest to you that autism can help us to understand that there are many forms of human consciousness as well. And this multiplicity may also account for the many conceptions of consciousness that I have reviewed with you.

But once again I must enlist your sympathy by reminding you of the impossibility of stepping outside of the system of consciousness, so that on this meta level we are all subject to mindblindness. And this brings to mind a story that is popular with general semanticists of all ages, the story of the blind men and the elephant. As you may recall, the story has one blind man touching the elephant's trunk and saying it's a snake, another touching its tail and saying it's a piece of rope, another touching a leg and saying it's a tree, and so on. Each one only senses part of the whole, and so they argue about what the true nature of the elephant really is. And for us, consciousness is the six ton elephant in our room, and there's no way to get him out of there. And it may be that the many conceptions of consciousness that I have reviewed with you are bits and pieces of some larger whole that we are unable to fully grasp, some phenomenon that exists in more dimensions than we can perceive. There is no way to know, of course, and so I return to the futility of my task, and recall the words of Ecclesiastes, "vanity, all is vanity." And perhaps I ought to give up at this point, but I am also reminded of the book title used by the renowned psychotherapist and communication theorist Paul Watzlawick: *The Situation is Hopeless But Not Serious* (1983). So let me now suggest some basic points.

First, I believe it is important to acknowledge that consciousness is not a thing. We think of it as a thing because the word "consciousness" is a noun. The fact that Indo-European languages such as English are especially noun oriented was made clear a long time ago by Benjamin Lee Whorf's (1956) study of the language of the Native American Hopi. As he put it, "Hopi, with its preferences for verbs, as contrasted to our own liking for nouns, perpetually turns our propositions about things into propositions about events" (p. 63). So, for example, our word lightning is a noun, and essentially we think of lightning as a thing, whereas the Hopi would speak of lightning as a verb and an event. And they would be right. Lightning is not an object or substance, it is a dynamic process, the movement of electrical energy across the sky. But if you have the mistaken idea that lightning is a thing, you might just conclude that lightning can be captured, controlled, and manipulated. And you might just get some old fool like Benjamin Franklin going out in a thunderstorm to catch lightning in a bottle. And that bottle becomes Thomas Edison's lightbulb, and John Ambrose Fleming's vacuum tube, and Jack Kilby and Robert Noyce's silicon computer chip. Sometimes the wrong map can take us into important new territory, as Christopher Columbus could tell you.

On a personal note, one evening when my son was about three years old, he looked out the window and said to me, "it's darking out!" And I said to myself, "Oh my God, my son's a Hopi." But once the initial shock wore off, I realized that his linguistic creativity brought him closer to reality than my own use of nouns such as "dusk" and "evening". Languages give us maps to guide us, but they do not dictate how we are to navigate. Moreover, the maps can be redrawn through linguistic invention, which is exactly what general semanticists try to do through techniques such as the elimination of the verb "to be".

So let us redraw our mental maps by thinking of consciousness as something we do rather than something we have. Consider the words of Buckminster Fuller as he self-reflexively considered his own consciousness: "I live on Earth, at present, and I don't know what I am. I know that I am not a category. I am not a thing—a noun. I seem to be a verb" (Fuller, Agel, & Fiore, 1970, p. 1). So I ask you to keep in mind that consciousness is a verb, but at the same time to retain the option of treating consciousness as if it were a noun. There is a certain utility in mistaking it for a thing if we want to catch lightning in a bottle, or catch a glimpse of the burning bush.

Understanding consciousness as an event and a dynamic process rather than a steady state, it then becomes possible to view it as an *ecology of mind*, to use Gregory Bateson's (1972) happy phrase. This means that we can view it as a system, and according to the German sociologist Niklas Luhmann (1989, 2005), the interdependent parts that make up the system of consciousness are thoughts. In other words, we do not begin with a fully formed consciousness, but rather with individual thoughts. As our thoughts become fruitful and multiply, we lose mental equilibrium and move toward the edge of chaos, until consciousness as a system emerges. We begin by thinking, and as thoughts proliferate, we start to think about thinking. In this way, consciousness emerges and gives us the means by which we may manage, organize, and perhaps control our own thought processes.

In referring to thought here, I want to make it clear that I am not only talking about rational ideas. Instead, I take as my authority the great philosopher of symbolic form, Susanne Langer (1957), who argued that human symbolic activity has as much if not more to do with feeling than with reason. For Langer, thought is not just a product of discursive symbols such as words, but also presentational symbols such as pictures. She would have recognized that Temple Grandin engages in a different but equally valid type of symbolic activity in her thought processes. Moreover, Langer explained that presentational forms begin with the senses, that the activity of perception is also a symbolic activity. In general semantics we understand that perception is the first rung up on the abstraction ladder, that we abstract or pull out of the world some of the sensory data from all that is available out there (Hayakawa & Hayakawa, 1990; Johnson, 1946; Korzybksi, 1993). And then our nervous systems process the information and create mental images, theories, or maps of the world. Through this active process we construct a view of the world as orderly, stable, consistent, predictable. This view is pure illusion, of course, reality being a chaos that we cannot live with, at least not and retain our sanity. So our senses filter out a large part of the data we take in, and the part that remains stands for the whole. Out of that limited form of chaos we construct a worldview. This parallels the emergence of consciousness from chaotic thought patterns.

Of course, we do not actually process data from the outside world. Rather, outside stimulation excites or irritates various nerve endings, causing them to transmit electrochemical signals through the system and on up to the brain. The process is the same when we register internal sensations such as a headache, back pain, or leg cramp. While we tend to speak of the five senses of sight, hearing, smell, taste, and touch, biologists also note the presence of such internal senses as the proprioceptive, which tells us about the movement and position of our joints and muscles, and the vestibular sense, our sense of balance, of gravity, of movement and position in relation to the earth. In fact, without proper development of the internal, intensional forms of perception, we have trouble developing our external senses, our thought processes, and ultimately our consciousnesses. The dichotomy between mind and senses, thought and perception, logic and empiricism is a false one. In the end, external perception and internal feeling, reason and emotion, all are based on neural activity, on the nervous system as information processor.

The nervous system is close to being a closed system, and it requires that closure in order to allow a worldview and a consciousness to form. If it were any more open, too much chaos would come in, and consciousness would not be able to organize itself. By the same token, the nervous system is not, and could not be, an entirely closed system. It needs that limited amount of chaos to allow for growth and increasing complexity. And while our sense of being in direct contact with the outside world is a false consciousness, it is still true that our perceptions are a response to outside stimuli, a reflection of the world. Even if they are just shadows on the cave wall, they are directly and immediately connected to what they represent, objective reality. As systems theorists Maturana and Varela (1992) put it (echoing Korzybski, 1993), we are *structurally coupled* with the world.

Perception, as a form of thought, is part of what gives rise to consciousness, and it guarantees that consciousness is not an entirely closed system. That is why the situation is hopeless but not serious. Certainly, it is not so serious as modern philosophers made it out to be when they introduced the problems of subjectivity and solipsism. In the twentieth century, the emphasis shifted from the isolated individual to the disconnected group, and the problems of intersubjectivity and the social construction of reality. And we may indeed be stuck in a room with no exit, and with that six ton elephant to boot, but we can crack open a window. Our social constructions are structurally coupled to the outside world, at least if we let them be.

Social construction is a very important idea, however, because we do construct our views of the world, albeit not out of whole cloth, and we construct them as a group, not individually. We tend to look at consciousness from the inside out, as an individual, internal, intensional phenomenon. But the word "consciousness" implies that it is a property of the group not the individual, that is, a social phenomenon. The "con" in consciousness tells us that it is about shared knowledge, shared awareness. What accounts for this commonality?

Certainly, there must be an underlying basis in the coding of our DNA, which then manifests itself in our bodies and nervous systems. But biology alone does not account for the development of consciousness as a group phenomenon. Rather, it is through interaction with others, through communication that we are bound together in a community of consciousness. The formation and development of the individual consciousness is triggered and shaped by relationships with parents and other significant persons in our environment. I would go so far as to say that communication is what enables human consciousness to form in the first place. To return to the metaphor of fire, consciousness is not a matter of spontaneous combustion, it is a spark traveling from one mind to another, a torch passed from one generation to another. Or to use a more recently coined term, it is a "meme" replicating itself in one brain after another.[*]

A basic point in systems theory is that systems are in a sense self-reflexive. Systems can exist within larger systems, and those systems within systems that are larger still. It follows that the individual consciousness, which can be understood as a system of thoughts or ecology of mind, can also be seen as an interdependent part of a larger system, a group or collective consciousness if you will. This is not necessarily a spiritual concept, as collective consciousness corresponds to some extent with the concept of culture. But this theory of metamind, the idea that we are part of a larger whole, is difficult to accept because we remain inside the system, unable to get out. Moreover, western culture's emphasis on individualism leads us to think of ourselves as alone and unique in the world. But on the biological level, we all share the same basic genetic code. And on the social level, members of the same culture use the same linguistic and symbolic codes. We think with forms of communication that are community property. We think with tools that are not of our own devising. As much as we would like to think otherwise, our thoughts are not ours alone.

Central to my understanding of consciousness is the counterintuitive idea that consciousness begins with communication. This means that consciousness begins externally, with interaction, and then becomes internalized over time. This process of the exterior becoming interiorized has been discussed by scholars such as George Herbert Mead (1934) and Walter Ong (1967, 1982, 2002). Of course, the process of internalization is largely an unconscious one, so that we are not aware and cannot recall the formation of our own conscious minds. But an example of internalization that you ought to remember occurs when we learn to read. Reading begins as an external process, as we sound out the letters to produce sounds, syllables, words, and sentences. After we learn to read out loud, we are then taught how to read silently, which is an internalization of reading. This is not an automatic development, and some children

---

[*] The concept of the "meme" was introduced, as a mental analogue to the biological notion of the gene, both being self-replicating entities, by biologist Richard Dawkins, in *The Selfish Gene*. (1989). Both meme and gene are logical constructs that are fairly high in order of abstraction, and often associated with essentialist orientations. Perhaps biologists need to remind themselves that the gene map is not the territory.

have great difficulty making the transition. Historically, silent reading was a rare and unusual practice in antiquity and the middle ages.

The internalization of reading and writing, I would suggest to you, parallels the interiorization of language that occurs much earlier in our development. The fundamental form of language is sound, and from the beginning we listen to the speech of others in our environment, and make sounds of our own. Typically, this leads to speech acquisition. We learn language by speaking out loud and listening to others as they do the same. Only later do we internalize these voices as memory and other forms of thought. Talking to one's self, then, reflects a temporary or perhaps chronic lack of interiorization. On the other hand, auditory hallucinations may represent a misrecognition of the same process. In this respect, Julian Jaynes (1976) offers some fascinating speculations about the origins of consciousness. He suggests that the first stirrings of full consciousness, that is, the early interiorizations of speech, were mistakenly interpreted as voices from the outside world, communication from supernatural beings. When thought itself was brand new, no one knew what to make of it, and they therefore concluded that it must be a god or spirit that was speaking to them (see also the critique of Jaynes in Ong, 1982).

The same process of interiorization works on all levels of symbolic activity. Even perception is interiorized as imagination, as visualization by the mind's eye, and as mental maps, phantom pains from lost limbs, hallucinations, memories, and dreams. Especially significant is the fact that our relationships with others are interiorized as well. Eric Berne (1961), founder of transactional analysis, argued that we all have an internalized parent as part of our own psyche. Alternately, according to George Herbert Mead (1934), we internalize not only specific others such as our parents, but a generalized other through which we can try to see ourselves as others see us. In this way we gain a measure of self-reflexiveness, self-awareness, self-consciousness. We also can internalize others when we imitate them and learn how to play roles. From the perspective of Mead, Hugh Duncan (1962, 1968), and Erving Goffman (1959), the internalization of role-playing is the way that we form a sense of self, or rather selves, for they see us as the sum of the roles that we play, the selves that we put on. From this point of view, consciousness is an act we perform in the theater of the mind. Put another way, individual consciousness is an interiorization of collective consciousness, an internalization of our relationships with other, pre-existing consciousnesses.

The internalization of the external is the process by which human consciousness is formed, and by which it changes. We can also say that consciousness changes because it is a dynamic ecology of mind, one that adapts to changes in its environment. For the individual consciousness, there is a process of growth and development, a process that may continue over the entire course of our lives. Paralleling the change that occurs on the individual level, we can also say that collective consciousness changes, that human consciousness evolves, as Walter Ong, Julian Jaynes, and many others maintain. And for our purposes it is sufficient to note that it is through a process of interiorization that human consciousness evolves.

And if consciousness changes and evolves, we can say that it has a past, and a future. Some of you probably thought that I was never going to get to the point of my essay, the topic that Allen Flagg assigned me, and others of you may have forgotten what it was in the first place. And all I can say is that it is in the nature of consciousness to wander. But the meandering path I have taken has, I hope, led us into important new territory, and brought us to the point that we can consider the road ahead. The media ecology perspective has at times been associated with futurism, but we have to begin by acknowledging that we are blind to the future, that all that we can really know is the past, and perhaps the present. This is what McLuhan called the rearview mirror effect (see McLuhan & Fiore, 1967). So, once more I seek the kindness and forgiveness of strangers and friends alike, as I indulge in some speculation and extrapolation in the performance of a task that is hopeless but not serious.

First and foremost, I would say that if we want to understand the future of consciousness, we need to study the new ways that we communicate in the present. And, if we want to understand the present, we need to put it into historical context. At some time in the distant and unremembered past, we, or our evolutionary ancestors developed spoken language, and at some later point we interiorized speech, which became a new form of thought, and the basis of a new form of consciousness. We internalized speech, and we also internalized the speaker, the other human beings that we interact with. When writing was invented, at first it was just a means of recording the spoken word. But eventually it was interiorized, giving rise to a new form of consciousness, a literate consciousness. This is a point I made earlier, but let me expand upon it. Before writing, words existed only as energy, as dynamic, ephemeral sound. After writing, words became fixed, static objects, nouns instead of verbs. In this way, we stopped talking and thinking like Hopi, and became westerners.

As Eric Havelock (1963) put it, writing separates the knower from the known. As we put our words, and therefore our thoughts down on paper, they become separate from ourselves, distanced and frozen, available to be viewed and reviewed. We can examine our thoughts as if they belonged to someone else, we can actually study our own minds in this fashion. Writing leads to entirely new levels of self-consciousness, but as Freud argues, at the price of repressing other parts of our consciousness into the unconscious mind.

Literacy is isolating. When we listen, we listen all together, as a group, but when we read, we read alone, even if we all read the same text. Literacy disrupts our sense of tribal identity and leads us to see ourselves as individuals, unique, alien and alienated. We come to think of consciousness as a private affair, and lose touch with the concept of collective consciousness. Only a highly literate mind could conceive of consciousness as an isolated island of subjectivity, or a hopeless prisoner of solipsism, and it is no accident that these philosophical movements materialize after the printing revolution in Europe. Literate consciousness distances us from the human lifeworld, and from our fellow human beings. As writers and readers, we confront the paradox that in order to communicate in this fashion, we require solitude, isolation. Literate consciousness also involves internalizing new roles, creating new selves. As Ong (1971, 1982) liked to say, the writer's audience is always a fiction. Which is to say that the writer must construct a new kind of generalized other, the reader. Ong also notes that the reader's writer is equally fictional, as we construct our own sense of the writer that we are reading, and internalize the role of the writer for ourselves.

As literacy becomes fully interiorized, writing turns more and more inward, towards the exploration of the mind. This is what the development of the novel is about, the shift from telling stories about agents performing actions to the interior examination of the individual consciousness. And as Ong has noted, it is no accident that Freud's depth psychology follows the development of the novel, that Freud discovers the unconscious and analyzes the complexities of consciousness only after fiction writers have turned inward. Freud was a product of the high point of literate culture, the moment when it was about to be overthrown by the new electronic media.

And is it any wonder that Korzybski was introducing his non-Aristotelian system at just about the time that radio and film had come to dominate our collective consciousness? Aristotle was very much a product of the old, literate consciousness, which is also characterized by highly abstract thinking. His teacher, Plato, was one of the first to think in so highly an abstract a manner. And in the same way that the first thoughts were mistakenly thought to be voices from beyond, the first high-level abstractions were mistakenly thought to be more real than our everyday reality. Plato was unable to recognize the figments of his own literate imagination, and that is why he arrived at his notion of ideal forms.

Moreover, as letter follows letter, word follows word, and line follows line, writing fosters a linear, one-step-at-a-time mode of thought. And this linear approach is the basis of Aristotle's logic. Also, his efforts at categorization are another example of the interiorization of writing's properties. Imagine that we began by making lists to keep track of things in our environment, for example the king's property. At a certain point, a single list become unmanageable, and someone says, let's break the list up, subdivide it into smaller, more specialized categories, so that one list is for livestock, another for pets for example. And this works just fine until someone says, well, this particular duck belongs in the category of livestock, but the king also treats it like a pet, so I will list it in both categories. And this double listing messes up our accounts, so someone else says, "If you list the duck twice, and someone else takes inventory, they'll think that there are two ducks, and that one of them is missing, and the king will think that we stole it and have our heads chopped off. So, no double listing, either you put an item on one list or on another, but not both!" From this concrete circumstance, we then interiorized the laws of noncontradiction and the excluded middle that are included in Aristotle's logic.

The future of consciousness, then, lies in the interiorization of contemporary communication technologies, specifically the electronic media. And the effects will be felt not only in what these media do, but, as Neil Postman (1992, 1995) liked to say, in what they undo.

For example, writing and especially printing have had a tendency to enforce a certain uniformity of mind, to go along with the uniformity of words fixed on a page, and the uniformity of the mass produced text. Our new media environment is characterized by a kaleidoscope of technologies, by a heterogeneous mix of oral, literate, visual, and tactile modes of communication and perception. This has been undoing the homogenizing effect of print media and, I would suggest, is opening the door to the coexistence of alternate modes of consciousness. This is reflected in Timothy Leary's experimentation with psychoactive drugs in the sixties, however dysfunctional that may have been, and in his embrace of virtual reality technology in the nineties. It is reflected in the popularity of transcendental meditation in the seventies, and the ongoing interest in New Age spirituality. It is also reflected in what many are now calling an epidemic of autism. While we parents look upon this with alarm, high functioning autistics object to the idea that their condition is a disease that needs to be cured, or even a disability that needs to be overcome. Instead they argue that theirs is a different mode of consciousness.

Although we may find a greater variety of consciousnesses in the future, not all will survive. Certainly, it seems that the tribal consciousness of oral cultures will continue on its march to extinction in the face of advancing technologies. But as some of literacy's effects are undone, we may see a return to a more oral-like consciousness, for example in the tendency to think less abstractly, more concretely and visually, and to become less distanced and objective, more emotionally involved with our world and our fellow human beings. I would also expect to see literacy's extreme individualism undone. We will not return to the simple group identity of oral culture, however. Instead, many of us now find ourselves in numerous relationships as we maintain contact with others through telephone, e-mail, instant messaging, and cell phones, in addition to more traditional modes of communication. We have more relationships with more individuals today than at any time in our history. This creates more selves for each of us, and a more fragmented and complex inner life. Postmodernists talk about the decentering of the subject, and psychologist Kenneth Gergen (1991) calls it the saturated self (see also Poster, 1990), but in another sense we are internalizing the heterogeneity of today's collective consciousness. The result is a less homogenous and uniform individual consciousness. I don't mean to suggest that the outcome will be some form of schizophrenia, but simply a new form of consciousness that reflects the chaos of the electronic media environment. It is a consciousness built upon constant, rapid stimulation occurring along multiple sensory channels simultaneously. McLuhan argued that the old linear mode of thought could not meet the demands of this new environment, and that we needed to develop new modes of pattern recognition to make order out of chaos. Pattern recognition, along with an internalization of the more recent phenomenon of multitasking, would be integral to a new type of electronic consciousness.

I also believe that the electronic media have removed many of the repressions that literate consciousness put into place. Certainly, we don't seem to be motivated by guilt or shame, at least not in the same way that members of literate and oral cultures were in the past. But in particular, I believe that the undoing of literate repression is putting us in closer touch with the unconscious mind both individually and collectively. In oral cultures, there was less of a barrier to the unconscious, allowing individuals to retrieve archetypes and enter the dreamtime with relative ease. But they did so in a fairly unreflective, unconscious manner. Having strengthened the conscious mind through literacy, we can now use the electronic media to engage the unconscious in a self-conscious manner. We are continually exposed to our dreams, and our nightmares, through television, film, and the internet. And we are discovering that there are many more monsters from the id out there than we ever suspected. But we are confronting them with full awareness. I therefore think it possible that we may find a way to integrate the conscious and unconscious minds, to embrace and absorb the shadow, the anima and animus, as Carl Jung (1969, 1978) referred to the components of the unconscious. Jung believed that such integration would leads us to the next stage in the evolution of consciousness, and it may be that this will happen as we interiorize our electronic communications. However one feels about Jung, it does seem that our survival as a species may depend on our ability to raise our consciousness to some higher level.

The evolution of consciousness is not a job for a lonely mystic isolated on a mountaintop. Rather we are all in this together, not just as passengers on spaceship earth and citizens of a global village, but as part of a collective consciousness more interconnected than ever before. We are rapidly approaching the point where we will be online all the time, and wherever we go. Within a generation, I believe that people will start to feel panic at the thought of going off the grid, of not being connected or having their vital signs monitored at every moment. Another generation, and that very possibility will become unimaginable. We seem to be moving toward a new form of collective consciousness. I don't think it is quite the same as Teilhard de Chardin's (1965) noösphere, although there does seem to be a spiritual dimension involved, at least as reflected in the renewed interest in spirituality in our time. But rather than a noösphere, it is a networked consciousness that we are creating, and internalizing as a new ecology of mind.

And this brings me to the end of my speculations and extrapolations, and the end of my essay about the future of consciousness. But at the end, I cannot help but introduce one last possibility, that the future of consciousness may be the end of consciousness. We may not be able to step outside of the system and fully imagine the end of all consciousness, both individual and collective, but we can imagine the possibility of the end. And having done so, we must be guided by that possibility. Alfred Korzybski saw the possibility of the end of consciousness when he was a soldier during the First World War. We here in New York City saw it just a few years ago, at the World Trade Center. And like Korzybski, we have to use the forms of consciousness that we have at our disposal today, to try to insure that there will still be forms of consciousness tomorrow.

References

Adams, C. (2003, September 19). How can a corporation be legally considered a person? Retrieved from The Straight Dope: http://www. straightdope.com/columns/read/2469/how-can-a-corporation-be-legally-considered-a-person

Always the etc.? (1949, February 16). *Time, 68*, 70.

Arcadia College. (n.d.). Retrieved from Wikipedia: http://en.wikipedia.org/wiki/Arcadia_University

Bardini, T. (2009/2010). When the map becomes the territory: Korzybski and cyberculture. *General Semantics Bulletin 76*, 37-49.

Barfield, O. *Poetic diction: A study in meaning* (2nd ed.). Middletown, CT: Wesleyan University Press.

Baron-Cohen, S. (1995). *Mindblindness: An essay on autism and theory of mind*. Cambridge, MA: MIT Press.

Bateson, G. (1972). *Steps to an ecology of mind: Collected essays in anthropology, psychiatry, evolution, and epistemology*. Chicago: University of Chicago Press.

Bateson, G. (2002). *Mind and nature: A necessary unity*. Cresskill, NJ: Hampton Press.

Baudrillard, J. (1981). *For a critique of the political economy of the sign* (C. Levin, Trans.). St. Louis, MO: Telos Press.

Baudrillard, J. (1983). *Simulations* (P. Foss, P. Patton & P. Beitchman, Trans.). New York: Semiotext(e).

Baudrillard, J. (1988). *The ecstasy of communication* (B. Schutze & C. Schutze, Trans.). New York: Semiotext(e).

Becker, E. (1971). *The denial of death*. New York: Free Press.

Becker, E. (1973). *The birth and death of meaning: An interdisciplinary perspective on the problem of man* (2nd ed.). New York: Free Press.

Berne, E. (1961). *Transactional analysis in psychotherapy: A systematic individual and social psychiatry*. New York: Grove Press.

Bertalanffy, L.v. (1969). *General system theory: Foundations, development, applications*. New York, G. Braziller.

Birdwhistell, R.L. (1970). *Kinesics and context: Essays on body motion communication*. Philadelphia: University of Pennsylvania Press.

Boulding, K. E. (1956). *The image: Knowledge in life and society*. Ann Arbor: University of Michigan Press.

Burke, K. (1950). *A rhetoric of motives*. Berkeley, CA: University of California Press.

Campbell, J. (1982). *Grammatical man: Information, entropy, language, and life*. New York: Simon & Schuster.

Campbell, J. (1968). *The hero with a thousand faces* (2nd ed.). Princeton, NJ: Princeton University Press.

Campbell, J. & Moyers, B.D. (1988). *The power of myth*. New York: Doubleday.

Capra, F. (1975). *The Tao of physics: An exploration of the parallels between modern physics and eastern mysticism*. New York: Shambhala.

Capra, F. (1982). *The turning point: Science, society, and the rising culture*. New York: Simon & Schuster.

Capra, F. (1996). *The web of life: A new scientific understanding of living systems*. New York: Anchor Books.

Capra, F. (2002). *The hidden connections: Integrating the biological, cognitive, and social dimensions of life into a science of sustainability*. New York: Doubleday.

Carey, J.W. (1989). *Communication as culture: Essays on media and society*. Boston: Unwin Hyman.

Carpenter, E. (1960). The new languages. In E. Carpenter & M. McLuhan (Eds.), *Explorations in communication* (pp. 162-179). Boston: Beacon Press.

Carpenter, E. & Heyman, K. (1970). *They became what they beheld*. New York: Outerbridge and Dienstfrey.

Carpenter, E. & McLuhan, M. (1960). *Explorations in communication*. Boston: Beacon Press.

Cassirer, E. (1953). *The philosophy of symbolic forms*. New Haven: Yale University Press.

Chase, S. (1938). *The tyranny of words*. New York: Harcourt, Brace.

Chomsky, N. (1972). *Language and mind*. New York: Harcourt Brace Jovanovich.

Cicero, M.T. (1972). *The nature of the gods* (H.C.P. McGregor, Trans.). Harmondsworth,England: Penguin.

Cosh, C. (2010, January 21). Le castor fait tout. Retrieved from http://www2.macleans.ca/2010/01/21/le-castor-fait-tout/#more-102965

Cozolino, L. (2006). *The neuroscience of human relationships: Attachment and the developingsocial brain*. New York: Norton.

Dawkins, R. (1989). *The selfish gene*. London: Oxford University Press.

Debray, R. (1996). *Media manifestos: On the technological transmission of cultural forms* (E. Rauth,Trans.). New York: Verso.

Debray, R. (2000). *Transmitting culture* (E. Rauth, Trans.). New York: Columbia University Press.

Drucker, P.F. (1946). *The concept of the corporation*. New York: John Day.

Duncan, H.D. (1962). *Communication and social order*. New York: Bedminster Press.

Duncan, H.D. (1968). *Symbols in society*. New York: Oxford University Press.

Eagleton, T. (1991). *Ideology: An introduction*. London: Verso.

Eisenstein, E.L. (1979). *The printing press as an agent of change:* Communications and cultural *transformations in early modern Europe* (2 vols.). New York: Cambridge University Press.

Einstein, A. (1935, May 4). The late Emmy Noether. [Letter to the editor]. *New York Times*, p. 12.

On the Binding Biases of Time

Einstein, A. (1983). *Sidelights on relativity*. New York: Dover.

Eliade, M. (1959). *The sacred and the profane* (W. Trask, Trans.). New York: Harvest/HBJ Books.

Eliade, M. (1975). *Myth and reality* (W. Trask, Trans.). New York: Harper Colophon Books.

Ellul, J. (1964). *The technological society* (J. Wilkinson, Trans.). New York: Knopf.

Ellul, J. (1965). *Propaganda: The formation of men's attitudes* (K. Kellen & J. Lerner, Trans.). New York: Vintage.

Ellul, J. (1985). *The humiliation of the word.* (J.M. Hanks, Trans.). Grand Rapids, MI: Williams B. Eerdmans.

Elson, L. (2010). *Paradox lost: A cross-contextual definition of levels of abstraction* (A. Ponikvar, Ed.). Cresskill, NJ: Hampton Press.

Emerson, R.W. (1883). *The conduct of life and society and solitude*. London: Macmillan and Co.

Epstein, S. & Epstein, B. (1964). *What's behind the word?* New York: Scholastic. Adapted from *The first book of words,* original work published 1954

Esté, A. (1997). *Cultura replicante: El orden semiocentrista*. Barcelona: Editorial Gedisa.

Ewen, S. (1988). *All consuming images: The politics of style in contemporary culture*. New York: Basic Books.

Forsdale, L. (1981). *Perspectives on communication*. Reading, MA: Addison-Wesley.

Fraser, J.T. (1987). *Time, The familiar stranger*. Amherst: University of Massachusetts Press.

Freire, P. (1970). *Pedagogy of the oppressed*. New York: Herder & Herder.

Frisch, M. (1959). *Homor faber: A report* (M. Bulloock, Trans.). San Diego: Harcourt BraceJovanovich.

Frith, U. (1989). *Autism: Explaining the enigma*. Oxford: Blackwell.

Freud, S. (1977). *Introductory lectures on psychoanalysis* (J. Strachey, Trans.). New York: Norton.

Frome, K.W. (2005). *What not to expect: A meditation on the spirituality of parenting*. New York: Crossroad Pub.

Fuller, R.B. (1971). *Operating manual for spaceship earth*. New York: E.P. Dutton.

Fuller, R.B., Agel, J., & Fiore, Q. (1970). *I seem to be a verb*. New York: Bantam Books.

Fuller, R.B. & Applewhite, E.J. (1975). *Synergetics: Explorations in the geometry of thinking*. New York: Macmillan.

Gardner, H. (1983). *Frames of mind: The theory of multiple intelligences*. New York: BasicBooks.

Gardner, H. (1993). *Multiple intelligences: The theory in practice*. New York: BasicBooks.

Gencarelli, T.F. (2000). The intellectual roots of media ecology in the thought and work of Neil Postman. *The New Jersey Journal of Communication 8*(1), 91-103.

Gencarelli, T.F. (2006). Neil Postman and the rise of media ecology. In C.M.K. Lum (Ed.), *Perspectives on culture, technology, and communication: The media ecology tradition* (pp. 201-253). Cresskill, NJ: Hampton Press.

Gencarelli, T.F., Borisoff, D., Chesebro, J.W., Drucker, S., Hahn, D.F., & Postman, N. (2001). Composing an academic life: A symposium. *The Speech Communication Annual 15*, 114-136.

Gergen, K. (1991). *The saturated self: Dilemmas of identity in contemporary life*. New York: BasicBooks.

Goffman, E. (1959). *The presentation of self in everyday life*. Garden City: Anchor Books.

Goffman, E. (1961). *Asylums: Essays on the social situation of mental patients and other inmates*. Garden City: Anchor Books.

Goffman, E. (1963). *Behavior in public places: Notes on the social organization of gatherings*. New York: Free Press.

On the Binding Biases of Time

Goffman, E. (1967). *Interaction ritual: Essays on face-to-face behavior.* Garden City: Anchor Books.

Goody, J. (1977). *The domestication of the savage mind.* Cambridge: Cambridge University Press.

Goody, J. (1986). *The logic of writing and the organization of society.* Cambridge: Cambridge University Press.

Goody, J. (1987). *The interface between the written and the oral.* Cambridge: Cambridge University Press.

Goody, J. (2000). *The power of the written tradition.* Cambridge: Cambridge University Press.

Goody, J. & Watt, I. (1968). The consequences of literacy. In J. Goody (Ed.), *Literacy in traditional societies* (pp. 27-68). New York: Cambridge University Press.

Gopnik, A., Meltzoff, A.N., & Kuhl, P.K. (2000). *The scientist in the crib: What early learning tells us about the mind.* New York: HarperCollins.

Gozzi, R. Jr. (1990). *New words and a changing American culture.* Columbia, SC: University of South Carolina Press.

Gozzi, R. Jr. (1999). *The power of metaphor in the age of electronic media.* Cresskill, NJ: Hampton Press.

Grandin, T. (1995). *Thinking in pictures and other reports from my life with autism.* New York: Random House.

Habermas, J. (1984-1987). *The theory of communicative action* (2 vols., T. McCarthy, Trans.).Boston: Beacon Press.

Hall, E.T. (1959). *The silent language.* Garden City: Doubleday.

Hall, E.T. (1966). *The hidden dimension.* Garden City: Doubleday.

Hall, E.T. (1983). *The dance of life: The other dimension of time.* Garden City: Anchor Press.

Havelock, E.A. (1963). *Preface to Plato.* Cambridge, MA: The Belknap Press of Harvard University Press.

Havelock, E.A. (1976). *Origins of western literacy*. Toronto: The Ontario Institute for Studies in Education.

Havelock, E.A. (1978). *The Greek concept of justice: From its shadow in Homer to its substance in Plato*. Cambridge, MA: Harvard University Press.

Havelock, E.A. (1982). *The literate revolution in Greece and its cultural consequences*. Princeton, NJ: Princeton University Press.

Havelock, E.A. (1986). *The muse learns to write: Reflections on orality and literacy from antiquity to the present*. New Haven, CT: Yale University Press.

Hawking, S.W. (1998). *A brief history of time* (rev. ed.). New York: Bantam Books.

Hayakawa, S.I. & Hayakawa, A.R. (1990). *Language in thought and action* (5th ed.). San Diego: Harcourt Brace.

Hayles, N.K. (1999). *How we became posthuman: Virtual bodies in cybernetics, literature, and informatics*. Chicago: University of Chicago Press.

Hill, T. The imagination in mathematics. *North American Review, 85,* 223-237.

Hobart, M.E., & Schiffman, Z.S. (1998). *Information ages: Literacy, numeracy, and the computer revolution*. Baltimore: John Hopkins University Press.

Hofstadter, D.R. (1979). *Gödel, Escher, Bach: An eternal golden braid*. New York: Basic Books.

Hughes, P. (1983). *More on oxymoron*. Harmondsworth, Middlesex, England: Penguin Books.

Innis, H.A. (1951). *The bias of communication*. Toronto: University of Toronto Press.

Innis, H.A. (1952). *Changing concepts of time*. Lanham, MD: Rowman & Littlefield.

On the Binding Biases of Time

Innis, H.A. (1972). *Empire and communications* (rev. ed., M. Q. Innis, Ed.). Toronto: University of Toronto Press.

Jameson, F. (1991). *Postmodernism, or the cultural logic of late capitalism.* Durham, NC: Duke University Press.

Jaynes, J. (1976). *The origin of consciousness in the breakdown of the bicameral mind.* Boston:Houghton Mifflin.

Jensen, J. (1990). *Redeeming modernity: Contradictions in media criticism.* Newbury Park, CA: Sage.

Johnson, M. (1987). *The body in the mind: The bodily basis of meaning, imagination, and reason.* Chicago: University of Chicago Press.

Johnson, W. (1946). *People in quandaries: The semantics of personal adjustment.* New York: Harper & Row.

Jones, S. (1992). *Rock formation: Music, technology, and mass communication.* Newbury Park, CA: Sage.

Jung, C.G. (1969). *The archetypes and the collective unconscious* (2nd ed., R.F.C. Hull, Trans.). Princeton: Princeton University Press.

Jung, C.G. (1978). *Aion: Researches into the phenomenology of the self* (2nd ed., R.F.C. Hull,Trans.). Princeton: Princeton University Press.

Kirk, G.S. (1962). *The songs of Homer.* Cambridge: Cambridge University Press, 1962.

Klein, W. (2010, January 26). Supreme Court ruling spurs corporation to run for Congress: First test of 'corporate personhood' in politics. *The Huffington Post.* Retrieved from http://www.huffingtonpost.com/ william-klein/supreme-court-ruling-spur_b_437871.html

Kodish, S.B. & Kodish, B.I. (2001). *Drive yourself sane: Using the uncommon sense of generalsemantics* (2nd ed.). Pasadena, CA: Extensional Pub.

Korten, D. (1999). *The post-corporate world: Life after capiralism.* San Fransisco: Berrett-Koehler.

Korzybski, A. (1950). *Manhood of humanity* (2nd ed.). Lakeville, CT: The International Non-Aristotelian Library/Institute of General Semantics. Original work published 1921

Korzybski, A. (1990). *Collected writings, 1920-1950*. Englewood, NJ: Institute of General Semantics.

Korzybski, A. (1993). *Science and sanity: An introduction to non-Aristotelian systems and general semantics* (5th ed.). Englewood, NJ: The International Non-Aristotelian Library/Institute of General Semantics. Original work published 1933

Korzybski, A. (2002). *General semantics seminar 1937* (3rd ed.). Brooklyn, NY: Institute of General Semantics. Original work published 1937

Korzybski, A. (2010). *Selections from science and sanity* (2nd ed.). Fort Worth, TX: Institute of General semantics. Original work published 1948

Kuhn, T.S. (1996). *The structure of scientific revolutions* (3rd ed.). Chicago and London: University of Chicago Press.

Lakoff, G. (1987). *Women, fire, and dangerous things: What categories reveal about the mind*.Chicago: University of Chicago Press.

Lakoff, G. & Johnson, M. (1980). *Metaphors we live by*. Chicago: University of Chicago Press.

Lakoff, G. & Johnson, M. (1989). *More than cool reason: A field guide to poetic metaphor*. Chicago: University of Chicago Press.

Lakoff, G. & Johnson, M. (1999). *Philosophy in the flesh: The embodied mind and its challenge to Western thought*. New York: Basic Books.

Langer, S.K.K. (1953). *Feeling and form: A theory of art*. New York: Scribner.

Langer, S.K.K. (1957). *Philosophy in a new key: A study in the symbolism of reason, rite and art* (3rd ed.). Cambridge, MA: Harvard University Press.

On the Binding Biases of Time

Langer, S.K.K. (1967). *Mind: An essay on human feeling* (Vol. 1). Baltimore: Johns Hopkins Press.

Langer, S.K.K. (1972). *Mind: An essay on human feeling* (Vol. 2). Baltimore: Johns Hopkins Press.

Langer, S.K.K. (1982). *Mind: An essay on human feeling* (Vol. 3) Baltimore: Johns Hopkins Press.

Laszlo, E. (1972). *The systems view of the world: The natural philosophy of the new developments in the sciences.* New York: G. Braziller.

Laszlo, E. (1996). *The systems view of the world: A holistic vision for our time.* Cresskill, NJ: Hampton Press.

Lee, D. (1959). *Freedom and culture.* Englewood Cliffs, NJ: Prentice-Hall.

Lee, I.J. (1994). *Language habits in human affairs* (2nd ed., S. Berman, Ed.). Concord, CA: International Society for General Semantics.

Lévi-Strauss, C. (1967). *Structural anthropology* (C. Jacobson and B.G. Schoepf, Trans.). Garden City, NY: Anchor Books.

Lévi-Strauss, C. (1969). *The raw and the cooked* (J. Weightman & D. Weightman, Trans.). New York: Harper & Row.

Levinson, P. (2010, January 21). Why the Supreme Court decision allowing direct corporate spending on elections is correct. Retrieved from http://paullevinson.blogspot.com/2010/01/why-supreme-court-decision-allowing.html

Lithwick, D. (2010, January 21). The Pinocchio project: Watching as the Supreme Court turns a corporation into a real live boy. *Slate.* Retrieved from http://www.slate.com/id/2242208

Logan, R.K. (1997). *The fifth language: Learning a living in the computer age.* Toronto: Stoddard.

Logan, R.K. (2000). *The sixth language: Learning a living in the Internet age.* Toronto: Stoddard.

Logan, R.K. (2004). *The alphabet effect: A media ecology understanding of the making of western civilization.* Cresskill, NJ: Hampton Press.

Logan, R.K. (2007). *The extended mind: The emergence of language, the human mind, and culture*. Toronto: University of Toronto Press.

Lord, A.B. (1960). *The singer of tales*. Cambridge, MA: Harvard University Press.

Luhmann, N. (1982). *The differentiation of society* (S. Holmes & C. Larmore, Trans.). New York: Columbia University Press.

Luhmann, N. (1989). *Ecological communication* (J. Bednarz, Jr., Trans.). Chicago: University of Chicago Press.

Luhmann, N. (1990). *Essays on self-reference*. New York: Columbia University Press.

Luhmann, N. (1995). *Social systems* (J. Bednarz , Jr. with D. Baecker, rans.). Stanford: Stanford University Press.

Luhmann, N. (2000a). *Art as a social system* (E.M. Knodt, Trans.). Stanford: Stanford University Press.

Luhmann, N. (2000b). *The reality of the mass media* (K. Cross, Trans.). Stanford: Stanford University Press.

Lum, C.M.K. (2006). *Perspectives on culture, technology, and communication: The media ecology tradition*. Cresskill, NJ: Hampton Press.

Luria, A.R. (1981) *Language and cognition* (J.Y. Wertsch, Trans.). Washington, DC: V.H. Winston.

Lyotard, J.F. (1984). *The postmodern condition: A report on knowledge* (G. Bennington & B. Massumi, Trans.). Minneapolis, MN: University of Minnesota Press.

Manovich, L. (2001). *The language of new media*. Cambridge, MA: MIT Press.

Marx, K. & Engels, F. (2007). *The Communist manifesto*. Charleston, SC: BiblioBazaar. Original work published 1848

Maturana, H.R. & Varela, F.J. (1980). *Autopoiesis and cognition: The realization of the living*. Boston: D. Reidel.

On the Binding Biases of Time

Maturana, H.R. & Varela, F.J. (1992). *The tree of knowledge: The biological roots of human understanding* (revised ed., R. Paolucci, Trans.). Boston: Shambhala.

McKibben, B. (2006). *The age of missing information*. New York: Random House.

McLuhan, E. (1997). *The role of thunder in Finnegans Wake*. Toronto: University of Toronto Press.

McLuhan, E. (1998). *Electric language: Understanding the message*. New York: Buzz Books.

McLuhan, M. (1962). *The Gutenberg galaxy: The making of typographic man*. Toronto: University of Toronto Press.

McLuhan, M. (1966). Cybernation and culture. In C. R. Dechert (Ed.), *The social impact of cybernetics* (pp. 95-108). Notre Dame, IN: University of Notre Dame Press.

McLuhan, M. (1969). *The interior landscape: The literary criticism of Marshall McLuhan, 1943-1962* (E. McNamara, Ed.). New York: McGraw-Hill.

McLuhan, M. (1976). The violence of the media. *Canadian Forum*, 9-12.

McLuhan, M. (2003). *Understanding media: The extensions of man* (Critical Ed., W. T. Gordon, Ed.). Corte Madera, CA: Gingko Press. Original work published in 1964.

McLuhan, M. (2006). *The classical trivium: The place of Thomas Nashe in the learning of his time* (W.T. Gordon, Ed.). Corte Madera, CA: Gingko Press.

McLuhan, M. & Fiore, Q. (1967). *The medium is the massage: An inventory of effects*. Corte Madera, CA: Gingko Press.

McLuhan, M. & Fiore, Q. (1968). *War and peace in the global village: An inventory of some of the current spastic situations that could be eliminated by more feedforward*. Corte Madera, CA: Gingko Press.

McLuhan, M. & McLuhan, E. (1988). *Laws of media: The new science*. Toronto: University of Toronto Press.

McLuhan, M. & Parker, H. (1968). *Through the vanishing point: Space in poetry and painting*. New York: Harper & Row.

Mead, G.H. (1934). *Mind, self and society from the standpoint of a social behaviorist* (C.W. Morris, Ed.). Chicago: University of Chicago Press.

Mead, M. (1964). *Continuities in cultural evolution*. New Haven: Yale University Press.

Meyrowitz, J. (1985). *No sense of place: The impact of electronic media on social behavior*. New York: Oxford University Press

Moran, T.P. (2007/2008). Media ecology as general semantics writ large. *General Semantics Bulletin 74/75*, pp. 38-40.

Mosk, M. (2010, January 26). O'Connor calls Citizens United ruling 'a problem'. Retrieved from http://abcnews.go.com/Blotter/oconnor-citizens-united-ruling-problem/story?id=9668044

Mumford, L. (1934). *Technics and civilization*. New York: Harcourt Brace.

Mumford, L. (1967). *The myth of the machine: I. Technics and human development*. New York: Harcourt Brace and World.

Mumford, L. (1970). *The myth of the machine: II. The pentagon of power*. New York: Harcourt Brace Jovanovich.

Nystrom, C. (1973). *Towards a science of media ecology: The formulation of integrated conceptual paradigms for the study of human communication systems*. Unpublished Doctoral Dissertation, New York University.

Nystrom, C. (1987). Literacy as deviance. *ETC: A Review of General Semantics, 44*(2), 111-115.

Nystrom, C. (2006). Symbols, thought, and "reality": The contributions of Benjamin Lee Whorf and Susanne K. Langer to media ecology. In C.M.K. Lum (Ed.), *Perspectives on culture, technology, and communication: The media ecology tradition* (pp. 275-301). Cresskill, NJ: Hampton Press.

Ogden, C.K. & Richards, I.A. (1923). *The meaning of meaning: A study of the influence of language upon thought and of the science of symbolism*. New York: Harcourt, Brace.

On the Binding Biases of Time

Ong, W.J. (1962). *The barbarian within and other fugitive essays and studies*. New York: Macmillan.

Ong, W.J. (1967). *The presence of the word: Some prolegomena for cultural and religious history*. New Haven, CT: Yale University Press.

Ong, W.J. (1971). *Rhetoric, romance, and technology: Studies in the interaction of expression and culture*. Ithaca, NY: Cornell University Press.

Ong, W.J. (1977). *Interfaces of the word: Studies in the evolution of consciousness and culture.* Ithaca, NY: Cornell University Press.

Ong, W.J. (1982). *Orality and literacy: The technologizing of the word*. London: Routledge.

Ong, W.J. (2002). *An Ong reader: Challenges for further inquiry* (T. J. Farrell and P. A. Soukup, Eds.). Cresskill, NJ: Hampton Press.

Ontario Consultants on Religious Tolerance. (n.d.). The Ten Commandments: Many topics, viewpoints, and interpretations. Retrieved from http://www.religioustolerance.org/chr_10co.htm

Orwell, G. (1949). *1984*. New York: New American Library.

Paglia, C. (1990). *Sexual Personae: Art and decadence from Nefertiti to Emily Dickinson*. New Haven: Yale University Press.

Parry, M. (1971). *The making of Homeric verse: The collected papers of Milman Parry* (Adam Parry, Ed.). Oxford: Clarendon Press.

Peirce, C.S. (1991). *Peirce on signs: Writing on semiotic*. Chapel Hill: University of North Carolina Press.

Perkinson, H. (1995). *How things got better: Speech, writing, printing, and cultural change*. Westport, CT: Bergin & Garvey.

Poster, M. (1990). *The mode of information: Poststructuralism and social context*. Chicago: University of Chicago Press.

Postman, N. (1961). *Television and the teaching of English*. New York: Appleton-Century-Crofts.

Postman, N. (1966). *Exploring your language*. New York: Holt, Rinehart & Winston.

Postman, N. (1967). *Language and reality*. New York: Holt, Rinehart & Winston.

Postman, N. (1968), November 29). Growing up relevant. Address delivered at the 58th annual convention of the National Council of Teachers of English, Milwaukee, WI.

Postman, N. (1970). The reformed English curriculum. In A.C. Eurich (Ed.), *High school 1980: The shape of the future in American secondary education* (pp.160-168). New York: Pitman.

Postman, N. (1974). Media ecology: General semantics in the third millennium. *General Semantics Bulletin 41-43*, 74-78.

Postman, N. (1976). *Crazy talk, stupid talk*. New York: Delacorte.

Postman, N. (1979). *Teaching as a conserving activity*. New York: Delacorte.

Postman, N. (1980). TV news as metaphor. *ETC: A Review of General Semantics 37*(4), 321-328.

Postman, N. (1982). *The disappearance of childhood*. New York: Delacorte.

Postman, N. (1985). *Amusing ourselves to death: Public discourse in the age of show business*. New York: Viking.

Postman, N. (1988). *Conscientious objections: Stirring up trouble about language, technology, and education*. New York: Alfred A. Knopf.

Postman, N. (1990, October 11). Informing ourselves to death. Retrieved from: http://w2.eff.org/Net_culture/Criticisms/informing_ourselves_ to_death.paper.

Postman, N. (1992). *Technopoly: The surrender of culture to technology*. New York: Alfred A. Knopf.

Postman, N. (1995). *The end of education: Redefining the value of school*. New York: Alfred A. Knopf.

On the Binding Biases of Time

Postman, N. (1996). Cyberspace, shmyberspace. In L. Strate, R. Jacobson, & S.B. Gibson (Eds.), *Communication and cyberspace: Social interaction in an electronic environment* (pp. 379-382). Cresskill, NJ: Hampton Press.

Postman, N. (1999). *Building a bridge to the eighteenth century: How the past can improve our future.* New York: Alfred A. Knopf.

Postman, N. (2000). The humanism of media ecology. *Proceedings of the Media Ecology Association 1*, 10-16.

Postman, N. & Damon, H.C. (1965a). *The uses of language.* New York: Holt, Rinehart & Winston.

Postman, N. & Damon, H.C. (1965b). *The languages of discovery.* New York: Holt, Rinehart &Winston.

Postman, N. & Damon, H.C. 1965c). *Language and systems.* New York: Holt, Rinehart & Winston.

Postman, N., Morine, H., & Morine, G. (1963). *Discovering your language.* New York: Holt, Rinehart & Winston.

Postman, N., Nystrom, C., Strate, L., & Weingartner, C. (1987). *Myths, men, and beer: An analysis of beer commercials on broadcast television, 1987.* Falls Church, VA: American Automobile Association Foundation for Traffic Safety.

Postman, N. & Powers, S. (1992). *How to watch TV news.* New York: Penguin Books.

Postman, N. & Weingartner, C. (1966). *Linguistics: A revolution in teaching.* New York: Delta.

Postman, N. & Weingartner, C. (1969). *Teaching as a subversive activity.* New York: Delta.

Postman, N. & Weingartner, C. (1971). *The soft revolution: A student handbook for turning schools around.* New York: Delacorte.

Postman, N. & Weingartner, C. (1973). *The school book: For people who want to know what all the hollering is about.* New York: Delacorte.

Postman, N., Weingartner, C., & Moran, T.P. (Eds.). (1969). *Language in America.* New York: Pegasus.

rat haus. (n.d.). U.S. Supreme Court – SANTA CLARA COUNTY v. SOUTHERN PAC. R. CO., 118 U.S. (1886). Retrieved from http://www.ratical.org/corporations/SCvSPR1886.html

Richards, I.A. (1929). *Practical criticism: A study of literary judgment.* New York: Harcourt, Brace.

Richards, I.A. (1936). *The philosophy of rhetoric.* New York: Oxford University Press.

Rifkin, J. (1987). *Time wars: The primary conflict in human history.* New York: H. Holt.

Ruesch, J. & Bateson, G. (1951). *Communication: The social matrix of psychiatry.* New York: Norton.

Rushkoff, D. (1994). *Media virus! Hidden agendas in popular culture.* New York: Ballantine.

Rushkoff, D. (2006). *Screenagers: Lessons in chaos from digital kids.* Cresskill, NJ: Hampton Press.

Russell, B. (1943). *An outline of intellectual rubbish: A hilarious catalogue of organized and individual stupidity.* Girard, KS: Haldeman-Julius Publications.

Ryan, P. (1993). *Video mind, earth mind: Art, communications, and ecology.* New York: Peter Lang.

Sapir, E. (1921). *Language: An introduction to the study of speech.* New York: Harcourt Brace Jovanovich.

Saussure, F.d. (1983). *Course in general linguistics* (C. Bally & A. Sechehaye with A. Riedlinger, Eds., R. Harris. Trans.). LaSalle, IL: Open Court.

Shannon, C.E. & Weaver, W. (1949). *The mathematical theory of communication.* Urbana: University of Illinois Press.

Shippey, T.A. (2000). *J.R.R. Tolkien: Author of the century.* Boston: Houghton Mifflin.

On the Binding Biases of Time

Shlain, L. (1998). *The alphabet versus the goddess: The conflict between word and image*. New York: Viking

Stendhal. (2002). *The life of Henry Brulard* (J. Sturrock, Trans.). New York: New York Review of Books. Original work published 1890.

Strate, L. (1982). Media and the sense of smell. In G. Gumpert & R. Cathcart (Eds.), *Inter/media: Interpersonal communication in a media world* (2nd ed., pp. 400-411). New York: Oxford University Press.

Strate, L. (1986). Media and the sense of smell. In G. Gumpert & R. Cathcart (Eds.), *Inter/media: Interpersonal communication in a media world* (3rd ed., pp. 428-438). New York: Oxford University Press.

Strate, L. (1994). Post(modern)man, Or Neil Postman as a postmodernist. *ETC: A Review of General Semantics, 51*(2), 159-170.

Strate, L. (2003). Cybertime. In L. Strate, R. Jacobson, & S.G. Gibson (Eds.), *Communication and cyberspace: Social interaction in an electronic environment*, (2nd ed., pp. 361-387). Cresskill, NJ: Hampton Press.

Strate, L. (2006). *Echoes and reflections: On media ecology as a field of study*. Cresskill, NJ: Hampton Press.

Strate, L. (2007). The Ten Commandments, as told by God to Moses (a new translation and interpretation by Lance Strate). Retrieved from http://blogs.myspace.com/index.cfm?fuseaction=blog.view&friendId=176504380&blogId=291115924

Teilhard de Chardin, P. (1965). *The phenomenon of man* (B. Wall, Trans.). New York, Harper and Row.

Theall, D.F. (1995). *Beyond the word: Reconstructing sense in the Joyce era of technology, culture, and communication*.Toronto: University of Toronto Press.

Theall, D.F. (1997). *James Joyce's techno-poetics*.Toronto: University of Toronto Press.

Tolkien, J.R.R. (1965a). *The fellowship of the ring*. New York: Ballantine.

Tolkien, J.R.R. (1965b). *The hobbit, or there and back again*. New York: Ballantine.

Tolkien, J.R.R. (1965c). *The return of the king*. New York: Ballantine.

Tolkien, J.R.R. (1965d). *The two towers*. New York: Ballantine.

Tolkien, J.R.R. (Ed. & Trans.). (1975). *Sir Gawain and the green knight, Pearl, Sir Orfeo* (C. Tolkien, Ed.). Boston: Houghton Mifflin.

Tolkien, J.R.R. (1977). *The silmarillion* (C. Tolkien, Ed.). New York: Ballantine.

Tolkien, J.R.R. (1980). *Unfinished tales of Númenor and Middle-earth* C. Tolkien, Ed.). Boston: Houghton Mifflin.

Tolkien, J.R.R. (1981). *The letters of J.R.R. Tolkien* (H. Carpenter with C. Tolkien, Eds.). Boston: Houghton Mifflin.

Tolkien, J.R.R. (1983a). *The book of lost tales: Part One* (C. Tolkien, Ed.). Boston: Houghton Mifflin.

Tolkien, J.R.R. (1983b). *The monsters and the critics and other essays* (C. Tolkien, Ed.). Boston: Houghton Mifflin.

Tolkien, J.R.R. (1984). *The book of lost tales: Part Two* (C. Tolkien, Ed.). Boston: Houghton Mifflin.

Tolkien, J.R.R. (1985). *The lays of Beleriand* (C. Tolkien, Ed.). Boston: Houghton Mifflin.

Tolkien, J.R.R. (1986). *The shaping of Middle-earth: The Quenta, the Ambarkanta, and the Annals* (C. Tolkien, Ed.). Boston: Houghton Mifflin.

Tolkien, J.R.R. (1987). *The lost road and other writings* (C. Tolkien, Ed.). Boston: Houghton Mifflin.

Tolkien, J.R.R. (1988). *The return of the shadow* (C. Tolkien, Ed.). Boston: Houghton Mifflin.

Tolkien, J.R.R. (1989). *The treason of Isengard* (C. Tolkien, Ed.). Boston: Houghton Mifflin.

Tolkien, J.R.R. (1990). *The war of the ring* (C. Tolkien, Ed.). Boston: Houghton Mifflin.

Tolkien, J.R.R. (1992). *Sauron defeated: The end of the third age, the Notion Club papers and the drowning of Anadune* (C. Tolkien, Ed.). Boston: Houghton Mifflin.

Tolkien, J.R.R. (1993). *Morgoth's ring: The later Silmarillion, part one* (C. Tolkien, Ed.). Boston: Houghton Mifflin.

Tolkien, J.R.R. (1994). *The war of the jewels: The later Silmarillion, part two* (C. Tolkien, Ed.). Boston: Houghton Mifflin.

Tolkien, J.R.R. (1996). *The peoples of Middle-earth* (C. Tolkien, Ed.). Boston: Houghton Mifflin.

Todt, R. (2000, November 20). Beaver College announces new name. Retrieved from http://abcnews.go.com/US/story?id=94962

UPI. (2010, January 12). Web forces Beaver magazine name change. Retrieved from http://www.upi.com/Odd_News/2010/01/12/Web-forces-Beaver-magazine-name-change/UPI-12051263323784

Veblen, T. (1921). *The engineers and the price system*. New York, B.W. Huebsch.

Virilio, P. (1986). *Speed and politics: An essay on dromology* (M. Polizzotti, Trans.). New York:Semiotext(e).

Virilio, P. & Lotringer, S. (1983). *Pure war* (M. Polizzotti, Trans.). New York: Semiotext(e).

Vonnegut, K. (1965). *God bless you, Mr. Rosewater, or pearls before swine*. New York, Delacorte Press.

Vygotsky, L.S. (1986). *Thought and language* (revised ed., A. Kozulin, Trans. & Ed.). Cambridge, MA: MIT Press.

Watzlawick, P. (1976). *How real is real?: Confusion, disinformation, communication*. New York: Random House.

Watzlawick, P. (1983). *The situation is hopeless, but not serious: (The pursuit of unhappiness)*. New York: Norton.

Watzlawick, P. (1988). *Ultra-solutions: Or, how to fail most successfully*. New York: Norton.

Watzlawick, P. (1990). *Münchhausen's pigtail: Or, psychotherapy & "reality": Essays and lectures*. New York: Norton.

Watzlawick, P., Bavelas, J.B., & Jackson, D.D. (1967). *Pragmatics of human communication: A study of interactional patterns, pathologies, and paradoxes*. New York: Norton.

Watzlawick, P., Weakland, J.H., & Fisch, R. (1974). *Change: Principles of problem formation and problem resolution*. New York: Norton.

Welcome to Oregon. (n.d.). Retrieved from http://www.el.com/to/oregon/facts

White, L. A. (1959). *The evolution of culture*. New York: McGraw Hill.

Whitehead, A.N. & Russell, B. (1925-1927). *Principia mathematica* (2nd ed., 3 vols.). Cambridge: The University Press.

Whorf, B.L. (1956). *Language, thought, and reality*. Cambridge, MA: MIT Press.

Wiener, N. (1950). *The human use of human beings: Cybernetics and society*. Boston: Houghton Mifflin.

Wiener, N. (1961). *Cybernetics: Or control and communication in the machine and animal*. Boston: Houghton Mifflin.

Williams, D. (1998). *Autism and sensing: The unlost instinct*. London: Jessica Kingsley.

Williams, R. (1977). *Marxism and literature*. Oxford: Oxford University Press.

Wittgenstein, L. (1961). *Tractatus logico-philosophicus* (D.F. Pears & B.F. McGuinness, Trans.). London: Routledge.

Wittgenstein, L. (1963). *Philosophical investigations* (G.E.M. Anscombe, Trans.). Oxford: Basil Blackwell.

Zingrone, F. (2001). *The media symplex: At the edge of meaning in the age of chaos*. Cresskill, NJ: Hampton Press.

Index

9/11, 158, 247

abstract, abstraction, abstracting 3, 5, 25-30, 32-33, 56-57, 59,-63, 103, 129-133, 148-152, 154, 175-177, 179, 181, 194, 211-212, 216, 221, 233, 238, 240, 244-245
Achilles, 182
acoustic space, 91, 168. See hearing, sound.
Adas Emuno, 11, 209
Adams, Cecil, 204, 251
Adler, Alfred, 35
adulthood, 20, 90, 107, 116
advertising, 90, 105, 116, 184-185, 213
aesthetic, 101, 104, 175-176, 179
Agel, Jerome, 55, 237, 255
aleph-bet, 145, 148
alphabet, 10, 45-46, 88-89, 260, 268
Alphaville, 35
American Museum of Natural History, 9
amputation, 61, 169. See also extension.
analogic, analogical, 106, 135, 178
analysis, 30, 89, 176
animal science, 235
animism, 147, 229
anthropology, 29, 55, 74, 82, 126-130, 134, 162, 212
Apollonian, 177
Apple Computer, 205
Applewhite, E.J., 53, 55, 255
Arcadia College, 191-192
archetypes, 89, 246
Aristotle, 5, 15, 19, 22-24, 40, 45-46, 48, 244
Aristotelian, 22-24, 29, 31-32, 40, 46, 133, 144, 178. See also non-Aristotelian.
art, 63, 95, 100, 101, 104, 126-127, 131-132, 176-179, 232-233
artificial intelligence, 229
Asimov, Isaac, 171
assembly line, 47
atomism, 30, 46. See also pre-Socratics.
audiovisual media, 104, 212

autism, autistic, 209, 234-235, 245
autopoiesis, autopoietic systems, 54, 56-57, 136

balance, 44, 84-87, 89-91, 94-96, 102, 104, 138, 186, 238. See also equilibrium, homeostasis, homeostatic system.
Bakshi, Ralph, 166
Bandler, Richard, 35
Bardini, Thierry, 11, 36, 251
Barfield, Owen, 161, 187, 251
Baron-Cohen, Simon, 229, 234, 251
Barthes, Roland, 118
Barwind, Jack, 10
Basic English, 124
Bateson, Gregory, 35, 55, 61-62, 81, 133-136, 139, 237, 251, 267
Baudrillard, Jean, 7, 36, 100, 102-103, 251
Bavelas, Janet B., 49, 55, 135, 163, 271
Beautiful Mind, A, 94
beaver, 9, 191-197
Beaver College, 191-192
Beavis and Butthead, 197
Becker, Ernest, 73, 158, 251
Berne, Eric, 241, 251
Bertalanffy, Ludwig von, 54-55, 61, 134, 227, 251-252
bias, 4, 6, 7, 43-48, 50, 60, 70-71, 80-91, 95-96, 106, 115-116, 124-125, 128, 131, 138, 180
Bible, 47, 67, 83, 143-154, 160
bifurcation, 53
Big Bang, 67, 96
binary opposition, 23, 107, 130, 135-136
biology, 18-19, 28, 35, 44, 53, 56, 58, 61, 71-72, 74, 134, 136, 185, 227, 231-233, 238-240. See also DNA, gene, life.
biotechnology, 48
Birds, The, 35
Birdwhistell, Ray L., 55, 252
Blade Runner, 94
book as medium, 4, 107

Boulding, Kenneth E., 55, 252
Bourne Identity, The, 94
brain, 69, 76, 130, 186, 214, 227, 230-231, 238, 240
broadcasting, 47, 90, 92
broadcast journalism, 117
Buber, Martin, 41
Buddhism, 36, 185, 229
Burke, Kenneth, 35, 44, 158, 252
Burns, Robert, 93
Burroughs, William S., 35
butterfly effect, 53
Buzan, Tony, 35

calendars, 67, 87-88, 152
Cambridge University, 161
Campbell, Jeremy, 136, 252
Campbell, Joseph, 158, 229, 252
capitalism, 19, 75, 78, 101, 103, 105, 201, 206. See also corporation.
Capra, Fritjof, 55, 134, 227, 252
Carey, James W., 41, 69-71, 81, 86, 90, 252
Carpenter, Edmund, 112, 115, 128-129, 137, 149, 162, 252-253
Cartesian coordinate graphing, 106
Cassandra, 40
Cassirer, Ernst, 131, 253
Catholic, Catholicism, Roman, 146, 157, 161, 171, 220
cell phones, 245
chaos, 40, 54, 61, 94, 147, 177, 229, 237-239, 246
Chase, Stuart, 34, 112, 132, 253
chat (online), 123
chemistry, 25-26, 44, 53-54, 58, 238
chemistry-binding, 18, 74, 76
child, children, childhood, 32, 93, 107, 116, 138, 217-224, 234, 240-241
Chomsky, Noam, 67-68, 130, 253
Christian, Christianity, 45, 83, 143, 171, 229
Cicero, Marcus Tullius, 86, 253
citizens, citizenship, 16, 46, 63, 118-119, 205, 246

Citizens United v. Federal Election Commission, 201
clocks, 55, 67, 69, 72, 87-88, 123
cloning, 48
cognition, 40, 157, 230, 232
cognitive science, 35, 230
coins, 46
Colorado State University, 235
Columbia University, 111, 162
Columbus, Christopher, 236
comic books, 171
command and control, 87
commemoration, 83, 212. See also memory.
commercialism, 19, 78, 89, 206, 213
communication, 3, 7, 9, 11, 22, 24-25, 28-29, 35-36, 41, 43, 47, 56, 58-60, 69-70, 79-96, 99-106, 111-115, 123-139, 149, 152-154, 162-163, 169, 175, 178, 180, 185-186, 209-210, 213-217, 231, 236, 239-241, 244-246. See also mass, communication.
communism, 19, 78
complexity, 31, 41, 54, 57, 74, 134, 136, 147, 239
computer, computing, 35, 47, 90, 92-94, 104-106, 117, 131, 135, 195, 219-220, 229-230, 236
computer-mediated communication, 91
consciousness, 8-9, 25, 29, 32, 36, 70-73, 81, 89, 96, 106, 124, 129, 131, 150-151, 154, 187-188, 220, 235-247
altered states of, 232-233, 235, 245
collective, 106, 239-240, 242-247
electronic, 244-247
false, 232, 239
historical, 171-172
literate, 242-246
networked, 247
of abstracting, 25-30, 32
of time, 71-73, 89, 96
raising, 29, 32, 150-151, 187, 232-233, 246
self-, 25, 131, 233-234, 241, 243
See also unconscious.

control, 29, 44, 69-70, 86-87, 92-93, 103, 106, 133, 154, 163, 215, 217, 236-237
Copernicus, Nicolaus, 42
constitution, 201-202, 206
construction, constructionism, constructivism. See social construction.
Cornell University, 10
corporation, 9, 201-206
Cosh, Colby, 193-197, 253
counter-environment, 116
Cozolino, Louis, 214, 253
crap detecting, 113
Cronkite, Walter, 203
cultural studies, 111
culture, 4-5, 7-8, 33-34, 43, 46, 68-70, 73-74, 81-94, 101-107, 115-119, 124, 128-130, 137-138, 148-151, 154, 162, 164, 167-168, 184-185, 196-197, 211-213, 218, 231, 240, 244-246
cummings, e e, 182-183
cuneiform, 62, 86, 88
cybernetics, 31, 35, 55, 133-134, 136, 138, 147, 229. See also feedback, thermostatic function, thermostatic view.
cyberspace, 117

Daily Show, The, 203
Damon, Howard C., 112, 266
dance, 132, 178
Dark City, 94
database, 91-93
dating. See extensional devices.
Dawkins, Richard, 71, 240, 253
DC Comics, 171
Dead Poets Society, 92
death, 67, 73, 94, 130, 158, 205-206, 228-230
Debray, Regis, 61, 124, 253
Decalogue. See Ten Commandments.
decontextualized, 103
Delaney, Samuel R., 35
democracy, 34, 46, 48, 101

Derrida, Jacques, 35, 118
Dewey, John, 36, 100
Dianetics, 35
Dick, Phillip K., 35
Dionysian, 177
difference, differentiation, 3, 18, 26, 32, 42, 44, 46-47, 50, 62, 79, 81, 94, 124, 126, 128, 134, 136, 147, 152, 229-230. See also order, organization.
digital, digitality, 47, 91-93, 106, 135, 178, 197, 215-216
dimension, 3-4, 23-24, 28, 68, 76, 128, 184, 210, 229, 236, 247
discourse, 102-106, 138, 204
discursive symbolism, 131, 135, 178, 237. See also proposition, propositional.
DNA, 74, 239. See also gene.
Douglas, William O., 201
dreamtime, 246
Drucker, Peter F., 205, 253
Duncan, Hugh D., 241, 253
Dungeons and Dragons, 172
durability, 87

e-mail, 123, 193, 245
E-Prime, 7, 237
Eagleton, Terrence, 232, 253
ecology, ecological, eco-logic, ecosystem, 3, 30, 39-44, 48-50, 58, 63, 95, 134, 138-139, 165, 217, 223. See also ecology of mind, media ecology, semantic ecology.
ecology of mind, 134, 237, 240, 242, 247
economics, 19-22, 34, 46, 70, 76-78, 80-81, 85-86, 95, 132, 136, 206
Edgerly, Mira, 15, 176
Edison, Thomas, 236
editing, 4-5, 8, 216
education, 7, 9, 19, 22, 34, 41, 46, 57, 99-100, 107, 111-118, 136-138, 162, 187, 209, 211, 215-218, 233. See also school, schooling; teaching.
effects, 45, 53, 55, 60, 89, 95, 125, 127
efficiency, 19, 60, 79, 90, 104, 106, 229

Eisenstein, Elizabeth L., 47, 57, 88, 101, 105, 136, 253

Einstein, Albert, 15, 17-19, 22, 30, 40-41, 54, 75-76, 83, 90, 113, 137, 147, 177-178, 253-254

ekphrasis, 177

Eliade, Mircea, 85, 254

electric, electricity, 31, 75, 90, 187, 219, 236

electrification, 78

electronic media, electronic media environment, electronic technology, 31, 47, 91-92, 95, 103, 105, 116-117, 129, 137, 168, 244-246

Ellul, Jacques, 47, 60, 100, 106, 254

Elson, Linda, 7-8, 123, 130, 138-139, 254

Elvish, 165, 172

emergence, 31, 53-54, 238

Emerson, Ralph Waldo, 61, 254

emotion, emotional, 29, 31, 33, 57, 131-132, 154, 157, 159, 161, 172, 179, 181, 186, 188, 238, 245

empathy, 234

empire, 69, 87, 89

empirical method, empiricism. See scientific method.

energy, 18-19, 23, 26, 54, 75-76, 79, 81, 83, 94, 236, 243

Engelbart, Douglas, 35

Engels, Friedrich, 75, 261

engineer, engineering, 15, 18, 20-22, 62, 69, 75, 78-81, 179, 182, 194. See also human engineering.

English,
    as an academic subject, 34, 62, 111-114, 124, 137-138, 157, 159, 162-163
    language, 7, 10, 15, 44, 45, 62, 67-68, 111, 128, 145, 172, 182, 236

Enlightenment, 46, 101, 118,

entropy, 40, 67, 94, 229

Ents, 165-168

environment, 3-4, 11, 18, 22, 28-29, 31, 36, 41, 44, 50, 54, 56-61, 63, 71-72, 74, 76, 79-81, 87, 95, 116, 135-136, 138, 153, 196, 210-212, 217, 220-221, 232, 239, 241-242, 244. See also counter-environment, media environment, semantic environment, symbolic environment.

epistemology, 29, 106

Epstein, Beryl, 10, 254

Epstein, Sam, 10, 254

equality, 46-47

equilibrium, 84, 133, 212, 237. See also balance, homeostasis, homeostatic system.

Esté, Achilles, 124, 254

etc. See extensional devices.

ETC: A Review of General Semantics, 5-9, 16-17, 36, 99, 114-155, 117, 119, 137

ethics, 20, 77-79, 95, 117, 143, 148-149, 182

Euclid, Euclidean, 40, 46, 106. See also non-Euclidean.

evaluation, 22, 25, 29, 44, 79, 113, 210, 212

event, 4, 24-29, 93, 131, 179, 193, 236-237

evolution, 9-10, 19, 44, 74, 78-79, 104, 151, 154, 231, 242, 246

Ewen, Stuart, 101, 254

excluded middle, 23, 178, 244

extension, media and technology as, 50, 61, 70, 73, 125, 169, 181, 206. See also amputation.

extensional devices, 7, 32-34, 193, 203
    dating, 7, 32
    etc, 32, 182-183
    indexing, 32, 193-196, 203

extensional orientation, extensionalism, 23, 31, 39, 154, 175, 179, 228, 233

exterior, 240

external, 26, 59, 61, 119, 220, 238-242

Facebook, 54, 91, 216, 219

fallibilism, 36

falsification, 42

feedback, 29, 53, 133, 137, 229. See also cybernetics, thermostatic function, thermostatic view.

film, 35, 92, 128-129, 158, 166, 171, 244, 246. See also motion pictures.

Fiore, Quentin, 55, 68, 163, 237, 242, 255, 262

fire, 75, 130, 231, 239-240

First Amendment, 201-204

Fisch, Richard, 49, 55, 135, 271

Flagg, Allen, 5, 10, 227, 242

FlashForward, 94

Flayhan, Donna, 6, 152, 219

Fleming, John Ambrose, 236

Flynt, Larry, 193

Fordham University, 5-6, 8-9, 11, 209, 220

form, 8, 25, 33, 55, 57, 59, 62, 63, 69-70, 83, 85, 89, 102-103, 106, 114, 116, 118-119, 123-139, 148-149, 152, 157, 162, 165, 171-172, 175-177, 179-181, 184, 186-188, 237-238, 244

Forsdale, Louis, 111-112, 128, 162, 254

Foucault, Michel, 36

Franklin, Benjamin, 75, 194, 236

Fraser, J.T., 71-72, 92, 254

Freire, Paolo, 215, 254

Frisch, Max, 61, 254

Frith, Uta, 229, 234, 254

Freud, Sigmund, 15, 31, 231-232, 243-244, 255

Frome, Keith W., 217, 255

Fuller, R. Buckminster, 35, 53, 55, 59, 237, 255

future, 4, 6, 9, 23, 68, 72-73, 81, 89, 92-96, 196-197, 205-206, 218, 227, 242, 244-247

Gaia hypothesis, 229

Galileo, 42

Gardner, Howard, 235, 255

Gencarelli, Thom F., 11, 115, 119, 255

gene, 71, 240. See also DNA.

general semantics, 3-11, 16-36, 39-44, 48-50, 53-54, 57-63, 70, 73-80, 90, 94-95, 111-115, 119, 123, 132-137, 144, 148, 151, 153-154, 162, 175-176, 179-186, 193-196, 203-204, 209-214, 216, 227-228, 233, 235, 237-238

General Semantics Bulletin, 6

general systems theory. See systems theory, systems view.

Gergen, Kenneth, 56, 245, 255

gestalt perception, 161-162

gestalt therapy, 35

global village, 58, 91, 103, 246

God, 41, 45, 86-87, 93, 143-144, 146-153, 171, 205-206

Goddess worship, 147

gods, 86, 146-147, 170, 229

Godard, Jean-Luc, 35

Gödel, Kurt, 25, 228

Goffman, Erving, 55-56, 114, 136, 241, 255-256

Goody, Jack, 44, 55, 83-84, 86, 133, 256

Gopnik, Alison, 214, 256

Gozzi, Raymond, Jr., 129, 256

Grandin, Temple, 235, 237, 256

graphical user interface (GUI), 35, 106

grammar, 3, 111, 128, 137, 162-163

Greece, Greeks, 39-40, 45-46, 48-49, 87, 168

GUI. See graphical user interface.

Gutenberg, Johannes, 47, 101-102, 172

Habermas, Jürgen, 153, 256

Haeckel, Ernst, 41

Hall, Edward T., 55, 61, 88, 114, 128, 136, 256

Hammurabi, 46

Hampton Press, 8

Havelock, Eric A., 45, 57, 82-83, 100, 114-115, 133, 136, 175, 243, 256-257

Hawking, Stephen W., 67, 257

Hayakawa, S.I., 34, 39-41, 111, 114-115, 119, 132, 162, 184-185, 195, 238, 257

Hayakawa, Alan R., 34, 132, 184, 238, 257

Hayles, N. Katherine, 56, 257

Healthy Media Choices, 9, 209-224

hearing, 62, 68, 150, 170-171, 238. See acoustic space, sound.

Hebrew, 45, 111, 145, 149

Hector, 182

Hegel, G.W.F., 102, 171

Heinlein, Robert, 35

Helen of Troy, 48

Heraclitus, 41, 46

Herbert, Frank, 35

heroes, 73, 158

Hesiod, 45

Heyman, K., 149, 252

hieroglyphics, 62, 86, 88

Hill, Thomas, 178, 257

Hindu, 46

historical consciousness. See consciousness, historical.

history, 5, 9, 21, 41, 67-69, 74, 79-80, 83-85, 92, 94-95, 101-103, 111, 136, 146, 160, 163, 168, 171, 229, 245

Hitchcock, Alfred, 35

Hobart, Michael E., 83, 257

Hobbit, The, 158, 165

Hofstadter, Douglas R., 35, 131, 257

homeostasis, homeostatic system, 49, 71, 74, 79, 84-86, 91, 95, 168. See also balance, equilibrium.

Homer, 45, 167, 182

Hopi language, 127, 236

Hubbard, L. Ron, 35

Hughes, Patrick, 227, 257

human, humanity, 3-4, 9, 17-22, 25, 28-29, 33, 41, 44, 50, 57-61, 71-74, 76-86, 90-91, 93-96, 115-117, 124, 126, 138, 147-150, 158, 162, 175, 178-180, 182-183, 188, 205-206, 210, 212, 214, 221, 231, 233, 237, 243, 245

human communication. See communication.

human consciousness. See consciousness.

human behavior, 7, 112, 124

human engineering, 16, 20-22, 78

human potential movement, 35-36

hyperrationality, 93

hyperreality, 93, 100, 103

hypertext, 93

identification, 26, 31, 44, 100, 184, 201-206

identity, 5, 23-24, 43-48, 133, 178

law of, 23, 31, 144

individual and collective, 81, 94, 192, 243, 245

ideology, 232

IFD disease, 48

image, 7, 33, 36, 62, 87, 92, 103, 106-107, 116, 119, 138, 146, 148-152, 197, 211, 221

incompleteness theorem, 25, 228

indexing. See extensional devices.

individual, individualism, 36, 43-44, 46-47, 58-59, 62, 91, 95, 126, 177, 187, 194, 237, 239-247

Indo-European languages, 236

industrialism, industrial revolution, 47, 101, 194

inference, 28, 33, 39, 146, 154, 228

infinite-valued orientation, 31, 126. See also multi-valued orientation.

information, 3, 42, 56-57, 59-60, 103-104, 117, 133-134, 136, 210, 218-220, 229-232, 234-235, 238

overload, 93, 103-105, 152, 188

society, 104

systems, 56-57, 229-232

theory, 133, 136, 147, 229

Innis, Harold A., 6, 41, 69-71, 80-91, 94-96, 100, 102, 115, 148, 171, 257-258

inquiry method, 113

instant messaging, 245

Institute for the History of Science, Polish

Academy of Sciences, 5

Institute of General Semantics, 10, 16, 209

intelligence, 131, 235

intensional orientation, 23, 154, 179, 228, 233, 238-239

interaction, 3, 41, 53, 60, 138, 211, 214, 239-240

interdependence, 41, 53

interface, 35, 59-60, 105-106

interiorization, interiorize, 43, 240-246

internalization, internalize, 9, 28, 59, 61, 119, 136, 220, 228, 238-247

International Society for General Semantics, 16, 111

International Society for the Study of Time, 71-72

internet, 91-92, 117, 123, 191-192, 195, 201, 215, 219, 222, 227, 246

interrelationship, 41

intersubjectivity, 239. See also social construction.

is, of identification, of predication, 31-32

Islam, 45, 143, 229

Israelites, 8, 45-46, 145-148

Jackson, Don D., 49, 55, 135, 163, 271

Jackson, Peter, 158-159, 166

James, William, 158

Jameson, Fredric, 7, 36, 80, 100-104, 258

Jaynes, Julian, 150, 241-242, 258

Jesuits, 157, 220

Jensen, Joli, 102, 258

Johnny Appleseed, 205

Johnson, Mark, 88, 129, 258, 259

Johnson, Nicholas, 35

Johnson, Wendell, 5, 35, 39-50, 95, 115, 133, 151, 162, 175, 238, 258

Johnston, Paul Dennithorne, 7

Jones, Steve, 93, 258

journalism, 35, 111, 117

Journeyman, 94

Judaism, 45, 83, 111, 143, 145, 149, 218-220, 223, 229

Jung, C.G., 246, 258

Kay, Alan, 35

Kendig, M., 17

Keyser, Cassius J., 15

Kilby, Jack, 236

Kilgore Trout, 205

Kirk, G.S., 46, 258

Klein, Jeremy, 7

Klein, William, 203, 258

knowledge, 10, 19, 21, 24, 26, 60, 63, 73, 76-77, 82-86, 88-92, 175, 180, 182, 186, 210, 212-213, 223

Kodish, Bruce I., 5, 18, 35, 258

Kodish, Susan B., 35, 258

Korten, David, 201, 258

Korzybski, Alfred, 3, 5-6, 10, 15-36, 39-41, 43-44, 47, 53-60, 62-63, 70-71, 73-82, 85, 90, 94-95, 100, 113, 115, 117, 119, 132-133, 137, 139, 162, 175-176, 179, 181-183, 206, 212, 239, 244, 247, 259

Kubrick, Stanley, 230

Kuhl, Patricia K., 214, 256

Kuhn, Thomas S., 75, 259

Lakoff, George, 88, 129, 259

Landman, Bette, 192

Langer, Susanne K.K., 41, 60, 131-133, 135, 139, 150, 157, 175, 237-238, 259-260

language, 3, 7-8, 10-11, 22, 24-26, 30-34, 36, 42-45, 48-49, 56-57, 60, 62-63, 67-68, 79, 85-86, 88, 96, 102, 111-113, 115-119, 124-139, 151, 162-167, 171-172, 175-182, 184-188, 194-196, 203, 210-212, 214, 233-237, 241-242

pollution, 113

Laocoön, 40

Laszlo, Ervin, 61, 134, 227, 260

Latin, 86, 88

laws of thermodynamics, 18, 67, 75

Leary, Timothy, 232, 245

Lee, Dorothy, 36, 62, 127-128, 162-163, 260

Lee, Irving J., 180, 260

Lennon, John, 93
Lévi-Strauss, Claude, 129, 212, 231, 260
Levinson, Paul, 11, 201, 260
Lewis, C.S., 157, 171
life, 18-20, 25, 40, 54, 60, 71-72, 74, 76, 80, 82, 96, 130, 144, 153, 158, 177, 180, 182, 205-206, 212, 228-231. See also biology.
Life on Mars, 94
linearity, 4, 30-31, 53, 68, 88, 92, 103, 133, 244, 246
linguistic, linguistics, 100, 103, 118, 125-127, 129, 157, 163-165, 175, 186, 188, 237, 240. See also philology.
linguistic relativism, 36, 162
Linnaeus, Carl, 19
literacy, 8, 55, 82, 87, 102-103, 128, 133, 148, 167-168, 197, 243, 245-246
literary theory and criticism, 35-36, 62, 89, 100, 124-129, 160-163, 171, 184-188
literature, 8, 6, 35, 126-127, 157-172, 175-188, 205-206, 243
Lithwick, Dahlia, 203, 260
Logan, Robert K., 11, 46, 56, 89, 128, 147-148, 260-261
logic, 5, 22-24, 31, 40, 45-46, 89, 94, 103, 105, 123, 130-131, 133, 148, 151, 176, 178, 204, 228, 238, 240, 244
logical types, theory of, 25, 130-131, 228
Long Now Foundation, 93
Lord, Albert B., 83, 261
Lord of the Rings, 157, 159-160, 164-166, 171-172
Lost, 94
Lotringer, Sylvere, 103, 270
Loyola, Ignatius, 220
Luhmann, Niklas, 5, 35, 56-57, 60-61, 63, 136, 237, 261
Lum, Casey Man Kong, 115, 255, 261, 263
Luria, Aleksandr R., 36, 130, 261
Lydia, 46
Lyotard, Jean-François, 7, 36, 100-101, 103-104, 261

magazines, 47, 123
Manovich, Lev, 129, 261
manuscript, 62, 88, 172. See also scribal, scribes.
map, 3, 24-25, 35, 50, 57, 59, 103, 153, 172, 177-179, 206, 210, 236-238, 241
map is not the territory, 24, 59, 63, 118, 179, 204, 210
Marvel Universe, 171
Marx, Groucho, 67, 75
Marxism, Marxist, 89, 99, 232. See also ideology.
Marx, Karl, 19, 100, 171, 261
mass, 18, 47, 182, 196
    communication, 47, 57, 81, 90, 123, 128, 136, 152, 213, 232, 245. See also communication.
    culture, 90, 101
    media. See mass, communication; media.
    persuasion, 90
    production, 47, 88, 245
    society, 47
mathematics, 15, 21, 33-34, 40, 46, 69, 80, 131, 135, 175-179
Matrix, 94
Maturana, Humberto R., 35, 56, 61, 136, 239, 261-262
McKibben, Bill, 221, 262
McLuhan, Eric, 11, 133, 161, 163, 262
McLuhan, Marshall, 5, 31, 36, 41, 44, 47, 53, 55-63, 68-70, 75, 89, 91-92, 100-104, 111-115, 119, 123-125, 128-129, 133, 137, 149, 161-163, 168-169, 172, 181, 197, 233, 242, 246, 252-253, 262-263
Mead, George Herbert, 100, 240-241, 263
Mead, Margaret, 63, 263
meaning, 5, 24, 29, 32, 40, 50, 67, 113-114, 116, 131-132, 135, 145, 175, 179, 184, 186, 193-194, 196, 211,

219, 234-235

mechanical reproduction, 197. See also mass communication, mass production.

mechanization, 47, 82, 88

media, 6, 9, 11, 43, 47-48, 50, 55, 57, 60-62, 70, 81, 85, 87-88, 91, 95, 99-100, 104-105, 111-119, 123-125, 128-132, 135-138, 148, 152, 161-163, 168-169, 172, 195, 197, 201, 206, 209-219, 232, 244-246

media ecology, 3-8, 10, 31, 36, 39-41, 43, 49, 53-57, 59, 61, 63, 70, 83, 100, 114-115, 119, 123-124, 126, 129, 131, 133-134, 136-139, 148-149, 154, 162-163, 167-168, 171-172, 197, 205-206, 209, 242

Media Ecology Association, 5-6, 119, 209

Media Ecology Book Series, 8

Media Ecology Graduate Program, 7-8, 10-11, 39, 114, 119, 138

media environment, 5-6, 8, 43, 45, 50, 56, 59-60, 116, 123, 138, 146, 154, 171, 197, 245-246

media literacy, 9, 215-218

media studies, 111, 209

mediation, mediating, 3, 5, 36, 56, 59-63, 81, 107, 124, 130

mediology, 61, 124

meditation, 152, 220-221, 235, 245

medium, 3-4, 6, 43, 45, 57, 59-61, 75, 79, 84, 88, 90-91, 115-116, 124-132, 135, 137, 139, 165-166, 175, 180, 186, 188, 212, 214, 220

medium is the message, 49-50, 58-59, 130, 132, 149, 163

medium theory, 56

Meltzoff, Andrew N., 214, 256

membrane, 56, 61

meme, 240

Memento, 94

memory, memorization, 46, 69-70, 72-73, 83-85, 91, 94, 153, 167, 179-180, 186-187, 212, 241. See also commemoration.

Menelaus, 48

Mental Research Institute, 135

metacommunication, 135

metalanguage, 180

metaphor, 4, 24, 68, 72, 74, 81, 86, 88-89, 125, 128-129, 162, 177-179, 195-196, 204, 210, 212, 220, 231-232, 239-240

metaphysics, 31-32, 128

metonymy, 60

Meyrowitz, Joshua, 56, 263

mimesis, 85, 184

mind, 31, 36, 40, 42, 57, 129-130, 134, 154, 161, 185, 188, 227-247

mindblindness, 234-235

mind mapping, 35

Minority Report, 94

mnemonic, 83, 175, 179-180, 186

Mobile telephones. See cell phones.

mode of abstracting, 62-63

modernity, 46, 80, 82, 101-107, 134, 162-163

Mohammed, 89, 143

money, 77, 88

monochronic, 88, 94

monomyth, 158

monopoly of knowledge, 88

monotheism, 45, 87, 146-151

Moran, Terrence P., 11, 39, 100, 113, 162, 263, 267

Morine, Greta, 112, 266

Morine, Harold, 112, 266

Morrison, Deborah, 192-193

MOSAIC (support group for autism), 209

Moses, 46, 143, 145, 154, 231

Mosk, Matthew, 201, 263

motion pictures, 35, 47, 62, 94, 104, 123, 197, 205, 235. See also film.

Moyers, Bill D., 229, 252

multimyth, 158

multitasking, 88, 246

multi-valued orientation, 31. See also infinite-valued orientation.

Mumford, Lewis, 36, 41, 47, 71, 78, 88, 100, 114, 171, 205, 263 music, 68, 72, 93, 131-132, 135, 170-171, 178

MySpace, 54, 91, 143

mystical experience, 233

myth, 69, 83-87, 92, 105, 130, 158, 165, 170-172, 197

nanosecond culture, 92

narrative, 73, 83, 92, 95, 103, 105, 157-158, 171-172

nation, nationalism, 58, 81, 194-195, 205-206

National Communication Association (formerly Speech Communication Association), 7

National Council of Teachers of English, 112, 114, 116, 137-138

neologism, 31, 129

NeoPoiesis Press, 209

nervous system, 25-28, 32, 59-60, 63, 92, 230-231, 233, 238-239

network, networks, 4, 54, 58, 91, 134, 161, 247

neural activity, 230-231, 238

neuro-linguistic programming (NLP), 35

neuroscience, 214

New Age, 228, 245

New Criticism, 125

New York Society for General Semantics, 5

New York State Communication Association, 6, 8, 209

New York University, 39, 111, 138

news, 89, 92, 101, 116-117, 138, 185, 191-193, 201

newspapers, 47, 62, 101, 123

Newton, Isaac, 42, 73

Newtonian, 23, 31, 40-41, 46

Nietzsche, Friedrich, 100, 195

non-allness, 24, 32, 133

non-Aristotelian, 3, 5-6, 10, 19, 22-25, 30, 32, 34-35, 41, 47, 133, 162, 175-176, 182, 193, 244

non-contradiction, 23, 178, 244

non-elementalism, 30-32, 57

non-Euclidean, 22, 40-41

non-identity, 24, 133, 178, 193

non-linearity, 31, 41, 55, 91, 134

non-Newtonian, 22, 40-41

non-verbal, 28, 33

nonverbal communication, 28, 33, 135, 221

noösphere, 247

notation, 85
  numerical, 46

novel, novelist, 35, 89, 157, 165-166, 172, 178, 205, 243

Noyce, Robert, 236

number, numbers, 40, 106-107, 119, 131, 135

Nystrom, Christine, 8, 11, 55-56, 100, 116, 138, 161, 263, 266

O'Connor, Sandra Day, 201

Odysseus, 39-41, 48-49

Ogden, C.K., 36, 61, 124-125, 263

Olivet College, 16

one-valued orientation, 31, 175

one cannot not communicate, 135

Ong, Walter J., 41, 44, 55, 57, 62, 68, 83-84, 91, 100, 102, 125, 133, 136, 150, 159, 168, 175, 197, 206, 240-243, 264

online, 70, 91-92, 123, 191, 203, 246

Ontario Consultants on Religious Tolerance, 143, 264

oral, culture, society, tradition; orality, 8, 45-46, 55, 63, 69-70, 82-92, 102, 127, 133, 138, 148, 167-168, 175, 185, 196, 212-213, 245-246. See also speech.

order, 26, 40, 44, 53, 61, 92, 147, 152, 177, 227, 229, 238, 246. See also difference, differentiation, organization.

organism, 30-31, 41, 58-61, 63, 71-72, 229-231

organization, 54, 70, 82, 136, 177, 229. See also difference, differentiation, order.

Organon (journal), 5

Orwell, George, 44, 116, 163, 178, 264
Oxford University, 157

paganism, 147, 229
Paglia, Camille, 176-177, 264
painting, 177-178
Palo Alto Group, 55, 135-136
paradox, 24, 102, 227-228, 243
Paradox Lost, 7, 131-139, 254
parenting, 209, 214, 217-218, 221-223,
    239, 241
Paris (of Troy), 48
Parker, Harley, 233, 263
Parry, Milman, 82, 167, 264
past, 4, 6, 23, 68, 72, 77-79, 81, 85-86,
    89, 92-94, 96, 188, 194, 219, 222-
    223, 242
pattern recognition, 246
Paul Bunyan, 205
Pavlov, Ivan, 15, 25, 29
Peirce, Charles Sanders, 124, 131-132,
    134, 264
perception, 3, 24-26, 28, 33, 60, 62, 72,
    79, 126, 131, 138, 151, 161-162,
    175, 181, 187, 210, 213, 233, 238-
    239, 241, 245
performance, 46, 83, 85-86, 93
Perkinson, Henry, 11, 89, 264
Perls, Fritz, 35
person, 201-206
personalism, 206
persuasion, 52, 90
Petkanas, Bill, 6, 8, 9-10
philology, 8, 157, 162-166, 168, 171.
    See also linguistics.
philosophy, 15, 22, 34, 100, 111, 114,
    131, 171, 206
Phoenicians, 45
photography, 104, 125, 196-197
physics, 17-18, 22-23, 40, 46, 69, 75,
    147, 178, 185
Plato, 20, 45, 176, 244
Pleasantville, 94
podcast, 9, 209, 215
poetry, 8-9, 33, 83, 91, 132, 143-144,

172, 175-188, 209
Poincaré, Henri, 15
Politechnika Warszawska, 15
political economy, 80
political science, 111
politics, 19-20, 46-47, 54, 57, 78, 89, 95,
    116, 136, 138, 182, 201, 203, 232
Ponikvar, Alan, 8, 123, 254
Popper, Karl, 36, 100
polychronic, 88, 94
polytheism, 87, 147
Poseidon, 39-40
positional numerical notation, 46
postal service, 123
Poster, Mark, 245, 264
Postman, Neil, 6-8, 10-11, 36, 39, 41,
    49, 55-56, 59, 90, 95, 99-107, 111-
    119, 125, 129, 137-139, 146, 148,
    162-163, 169, 197, 233, 244, 255,
    265-267
postmodern, postmodernism, 7, 36, 46,
    80, 90, 94, 99-107, 118, 187, 245
post-scientific, 80
Powers, Steve, 100, 106, 117, 266
pragmatism, 34, 100, 124, 132
prelinguistic, 235
Premonition, 94
presence, 147-148,
present, 6, 21, 47, 72, 84-85, 89-96, 99,
    101, 227, 242
present-centeredness, 92, 94, 103
presentational symbolism, 131, 135, 150,
    178, 237-238
preservation, 81, 87
pre-Socratics, 30, 45-46.   See also
    atomism.
print, printing 10, 47, 57, 88-89, 101-102,
    104-106, 116, 118, 137-138, 185,
    197, 243, 245. See also typography.
privacy, 103
profane space, 98, 158
profane time, 85, 94-95, 152
progressive religion, 9, 219-220, 223
progress, 10, 19-20, 73-74, 78-80, 84-85,
    89, 93, 95, 104, 180, 182, 194, 212,

219
proprioceptive, 238
proposition, propositional, 33-34, 70, 77, 103, 131, 146, 157, 236. See also discursive symbolism.
psychiatry, 15-16, 134
psychology, 15, 20, 25, 29, 35-36, 41, 53, 55, 61, 111, 129, 132, 134, 232, 243, 245
psychotherapy, 35, 92, 134-135, 236

quandaries, 48-49
quarrels, 48-49
quagmires, 48-49
questions, 25, 39-40, 42, 48-50, 113, 133, 181, 204-206

Raiders of the Lost Ark, 145
radio, 9, 62, 90, 104, 123, 128, 192, 209, 244
rat haus, 201-203, 267
rational emotive therapy, 35
Read, Charlotte Schuchardt, 17
reading, 4, 86, 88-89, 185-186, 197, 240-241, 243
reality, 3, 24-25, 28-29, 34, 39-41, 50, 57, 72, 91, 94, 106, 118, 124, 126, 128, 133, 135-136, 176-181, 204, 209-211, 232, 237-239, 244
reality-testing, 23
rearview mirror, 68, 242
recording, 47, 62, 83, 92-93, 123, 213, 242
recursion, 131, 181, 228
relationship, 3, 23, 30, 41, 50, 53, 58-59, 63, 70, 82, 106, 135, 186, 239, 241-242, 245. See also identity.
relativism, 90. See also linguistic relativism.
relativity, 19, 30, 40, 41, 54, 67, 76, 90, 177
religion, 45, 54, 69, 85-87, 89, 95, 116, 138, 143-154, 157-159, 161-162, 171, 211, 219-220, 229, 233
religious studies, 111

repression, 246
rhetoric, 35, 44, 74, 118, 129, 162
Richards, I.A., 36, 61, 100, 124-125, 186-187, 263, 267
Rifkin, Jeremy, 92, 267
ritual, 69, 85-87, 131, 149, 171, 218, 224
ritual view of communication, 81
Rogers, Carl, 41
role, 10, 166, 241, 243
Roman Catholic, Roman Catholicism. See Catholic, Catholicism, Roman.
Roosevelt, Franklin Delano, 34, 111
Rose, Ellen, 6
Royce, Josiah, 15
Ruesch, Jurgen, 134, 267
Rushkoff, Douglas, 11, 56, 228, 267
Russell, Bertrand, 15, 25, 33, 36, 130-131, 133, 139, 178, 228, 267, 271
Ryan, Paul, 124, 267

Sabbath, 94, 144-145, 151-152, 219-221
sacred space, 85, 158
sacred text, 45, 86, 144
sacred time, 85-86, 88, 94, 152
Saint Louis University, 159
San Francisco State College/University, 34, 111
sanity, 29, 42, 50, 90, 238
Santa Clara County v. Southern Pacific Railroad Company, 201-202
Santa Claus, 205
Sapir, Edward, 36, 62, 79, 126-127, 129, 162, 267
Sapir-Whorf hypothesis, 79, 128
Sapir-Whorf-Lee hypothesis, 128, 163
Sapir-Whorf-Korzybski-Ames-Einstein-Heisenberg-Wittgenstein-McLuhan-Et Al. hypothesis, 113, 137
Saussure, Ferdinand de, 100, 124, 267
Schiavo, Terry, 231
Schiffman, Zachary S., 83, 257
school, schooling, 15, 34, 92, 95, 103, 107, 112, 115-116, 137-138, 217-218, 220, 232. See also education, teaching.

science, 18, 20-23, 30, 33-35, 41-42, 44, 48-49, 69, 74-75, 77-80, 85, 89, 95, 103, 105, 147, 153, 175-176, 179, 185

science fiction, 35, 68, 171, 205

scientific method, 20, 22-23, 30-31, 34, 36, 39, 41-42, 79-80, 89, 100, 131, 175, 210, 228, 238

Scientology, 35

scribal, scribes, 88, 145, 172. See also manuscript.

sculpture, 132, 177-178

secondary orality, 91, 102, 168

Seger, Pete, 67

self, sense of, 107, 233-234, 241, 245. See also consciousness, self-.

self-esteem, 73

self-reference, self-referential. See self-reflexiveness.

self-reflexiveness, 24-25, 28, 131, 133-136, 145, 181, 227-228, 237, 240-241

self-replication, 74, 240

self-organization, 54, 60, 72, 136. See also autopoiesis, autopoietic systems.

semantic ecology, 43, 113-114

semantic environment, 8, 43, 50, 59, 113-114, 146, 154, 184. See also symbolic environment.

semantic reaction, 26, 29, 40, 57-58

semantics, 5, 29, 137. See also general semantics.

semiotics, 36, 116, 124, 134, 197

semiology, 61, 124

Semites, 45-46, 148

shamanism, 147

Shannon, Claude E., 133, 229, 267

Shippey, Thomas A., 157, 159-161, 163-165, 267

Shlain, Leonard, 147-148, 268

sight, 171, 238. See also vision, visualism.

signal, 152, 238

signal reaction, 29, 150

sign, 25, 39-40, 124-125, 157

signification, 3, 25, 61, 124, 152

signifier, 103, 124, 194

signified, 103, 124, 196

Silmarillion, 160, 165, 170, 269

simulation, 103, 106

slang, 196

smell, 62, 68-69, 150, 167, 238

social construction, 40, 118, 126, 135-136, 138, 232, 239. See also intersubjectivity.

social learning, 74

social media, 91, 215

social networking, 54, 91, 161

social organization, 70, 81-82, 85, 89, 117, 222

social systems, 5, 54, 56-57, 60-61, 136

socialism, 78

society, 19-22, 43, 46-47, 54, 78, 80, 85-87, 90, 94-95, 104, 129, 136, 152-153, 176, 180, 186-187. See also mass, society; social organization; social systems.

sociology, 101, 111

solipsism, 239, 243

sound, 4, 68, 82-83, 164, 170, 180, 235, 240-241, 243. See acoustic space, hearing.

space, 4, 30, 40, 68-71, 76, 80-82, 85, 87, 91, 95, 115, 158, 168, 188

space bias, 4, 69-70, 87-91

space-binding, 19, 69-70, 74, 76

spaceship earth, 59, 246

spacetime, 31, 76

speech, speech act, 35, 45-46, 48, 62, 129, 133, 153, 163-165, 171-172, 175, 177, 180, 186, 188, 204, 241-242. See also oral, culture, society, tradition; orality.

Speech Communication Association. See National Communication Association.

speed, 75, 87, 103

Spengler, Oswald, 171

Spider-Man, 205

spirituality, 9, 86-87, 92, 147, 149, 158, 188, 219-221, 229, 240, 245, 247

Stalin, 17, 178

Star Trek, 171

Star Wars, 171

State University of New York at Freedonia, 111

Stendhal, 178, 268s

stereotypes, 34, 94, 209, 211

Stewart, Jon, 203, 205

Strate, Lance, 9, 36, 41, 56, 91-93, 100, 119, 123, 143, 162, 197, 209, 266, 268

structural coupling, 239

structural differential, 27-28

structure, 3, 24-26, 30-31, 36, 43, 53, 67, 69-70, 115, 118, 124, 130, 138, 152, 210, 222

subjectivity, 25, 28, 33, 91, 239, 243

Superman, 205

Supreme Court, 9, 201-203,

surveillance, 91

survival, 18, 44, 71, 84, 138, 229, 246

symbol, symbolic, symbolization, symbol system, 3, 8, 17, 22, 24-25, 29, 39-41, 43-47, 50, 55, 57-60, 62-63, 69, 79, 112, 114-119, 123-126, 129-139, 149-150, 152, 157, 162, 169, 171-172, 175, 178-181, 184-185, 194-195, 197, 204, 210-212, 214, 221, 231-233, 237-238, 240-241

symbol reaction, 29, 151-154

symbolic environment, 59, 171. See also semantic environment.

synchronization, 87

synergy, 53

synthesis, 30

systems theory, systems view, 3, 5, 10-11, 31, 35, 49, 53-63, 72, 123, 129, 132-138, 147, 227-230, 235, 237-240, 247

Taoism, 36

taste, 238

Teachers College. See Columbia University.

teaching, 22, 28, 79, 99, 107, 112-113, 144, 215, 233. See also education, school, schooling.

technics, 115

technique, 43-44, 47, 60, 63

technocracy, technocratic, 20, 78, 104-105

technologization, 3

technology, 20, 34, 45, 48, 55-56, 60-61, 63, 69, 75, 78-80, 87, 89-90, 95, 99-100, 102-106, 111, 114, 116-119, 123, 125, 128, 130, 137, 150, 169, 188, 209, 213, 231, 245

technology studies, 119

technopoly, 90, 104-106, 117

Teilhard de Chardin, Pierre, 247, 268

telecommunications, 91-92, 103, 169

telegraph, 75, 90, 104, 123

telephone, 90, 123, 125, 245

television, televisual, 62, 91, 94, 102-106, 116, 123, 138, 160, 169, 171, 191, 212-213, 216-222, 246

Ten Commandments, 8, 143-154

Tennyson, Alfred Lord, 182

Terminator, The, 94

territory, 3, 24, 59, 63, 89, 101, 118, 153, 177, 179, 204, 206, 210, 236, 242

terrorism, 47, 158

text messaging, 123

Theall, Donald F., 163, 268

theory of mind, 229, 233-234

thermostatic function, thermostatic view, 55, 95, 137. See also cybernetics, feedback.

thinking, 29, 31, 39-43, 68, 74, 89, 134, 144, 148-149, 152, 161, 183, 185, 204, 205, 221, 227, 229, 235, 237, 243-244

thought, 4, 10, 17, 23-24, 31, 36, 41, 69, 73, 111, 126, 129, 132-133, 148, 150, 152-154, 161, 163-166, 170-171, 175, 179, 181, 188, 233, 237-244, 246

time, 4-5, 23-24, 30, 40, 67-96, 101-102, 151-152, 159, 166, 168, 188, 196, 212-213, 217-218, 220-222, 244, 246

time-binding, 4-5, 18-20, 22, 24, 57-58, 60, 69-71, 73-87, 91-94, 96, 153, 175, 180, 182, 186, 211-213, 216, 219
time bias, 4-5, 69-71, 80-91, 96, 115
time consciousness, 71-73
Time magazine, 16-17
time travel, 68, 94
Time-Warner, 205,
"to be", 31-32, 45, 237
Tolkien, Christopher, 160-161, 277-278
Tolkien, J.R.R., 8, 157-172, 277-278
Todt, Ron, 192, 270
Tom Sawyer, 205
total recall, 91-92
Total Recall, 94
touch, 62, 236, 238
Toynbee, Arnold, 171
traffic, 227
transcendental meditation, 245
transformation, 231
transportation, 87
transportation view of communication, 81
tribal, tribalism, 68, 74, 85-86, 91, 212, 243, 245
trivium, 128, 162-163
Trojan horse, 39-40
Truman Show, The, 94, 205
truth, 40, 128, 175, 179, 220, 232
Twitter, 91, 216
two-valued orientation, 31, 57
typography, 31, 89, 102-107, 197. See also print, printing.

uncertainty, 41, 90, 93, 228
unconscious, 31, 40, 186, 213, 230-232, 240, 243, 246
University of Chicago, 80
University of Illinois, 17
University of Maine, 6
University of Rome, 15
University of Toronto, 80
University of Washington, 214
universe, 18, 23, 40, 46, 50, 54, 67, 75, 94, 147, 159, 171, 229

universe of discourse, 204
UPI, 143, 191-192, 270
urbanization, 101

Van Vogt, A.E., 35
Varela, Francisco J., 35, 56, 61, 136, 239, 261-261
Veblen, Thorstein, 78, 270
verb orientation, 25, 59, 61, 236-237, 243
verbal communication, 7, 26, 28, 33, 59, 102-103, 106, 166, 178-179, 184
vestibular, 238
Virgil, 67
Virilio, Paul, 103, 270
virtual reality, 245
vision, visual, 4, 33, 35, 62, 68, 82-83, 102-103, 106, 148, 150, 166, 168-169, 171, 196-197, 235, 241, 245. See also sight.
Vonnegut, Kurt, 205, 270
Vygotsky, Lev S., 36, 129, 270

war, 21, 33, 44, 47, 68, 74, 206
    Cold, 47, 90
    First World War, 15, 47, 80, 90, 160, 182-183, 247
    of the Ring, 166-169
    on terror, 47
    Second World War, 16, 18, 47, 80, 90, 222
    Trojan, 39-41, 45, 47-48
Watts, Alan, 35
Watzlawick, Paul, 35, 41, 49, 55-56, 114, 135-137, 139, 163, 236, 270-271
Weakland, John H., 49, 55, 135, 271
weapons of mass destruction, 18, 47, 182
Weaver, Warren, 133, 229, 267
Weingartner, Charles, 36, 99-100, 112-113, 116, 129, 137-138, 233, 266-267
Wells, H.G., 171
White, Leslie A., 128, 136, 271
White, William Alanson, 15

On the Binding Biases of Time

Whitehead, Alfred North, 15, 25, 33, 36, 130-131, 133, 139, 228, 271

Whorf, Benjamin Lee, 36, 62, 79, 127, 162, 236, 271

Winslow, Dale, 8

WFUV, 9, 209

WVEW, 9

Wiener, Norbert, 54-55, 114, 133-134, 229, 271

Williams, Donna, 234-235, 271

Williams, Raymond, 232, 271

Williams, Robin, 92

Wilson, Robert Anton, 35

Witness for Childhood, 9, 219

Wittgenstein, Ludwig, 36, 138-139, 271

word, 10, 24, 26, 28, 32-34, 36, 43, 45, 48, 50, 62, 85, 99, 102-103, 106, 116, 119, 125, 131, 133, 138, 148-151, 164, 167, 177, 179-181, 185, 188, 193-194, 196, 221, 234-237, 240, 242-245

worldview, 40-41, 46, 79, 91, 104, 106, 126, 238

writing, 4, 7, 10, 24, 32, 44-46, 57, 62, 83, 85-89, 92, 102, 106, 145, 160, 167, 172, 175, 177, 180, 181, 185, 186, 197, 212, 235, 241-245

Yale University, 16-17, 82

Yiddish, 93, 111

Zaborowski, Robert, 5

Zingrone, Frank, 56, 129, 271

Lance Strate is Professor of Communication and Media Studies at Fordham University (1989-present), and has served as the Executive Director of the Institute of General Semantics (2008-2011). One of the founders of the Media Ecology Association, he was the MEA's first president (1998-2009), and remains a member of its Board of Directors; he also served as president of the New York State Communication Association (1998-1999), and is a member of the Board of Directors of the New York Society for General Semantics (2009-present). The author of *Echoes and Reflections: On Media Ecology as a Field of Study* (2006), and over 100 articles and book chapters, he co-edited *The Legacy of McLuhan* (2005), and two editions of *Communication and Cyberspace: Social Interaction in an Electronic Environment* (1996, 2003). He has been the editor of several journals including the *General Semantics Bulletin* (2007/2008), the *Speech Communication Annual* (2000-2001), and *Explorations in Media Ecology* (2002-2007) which he founded, as well as co-editor of the "Poetry Ring" feature for *ETC* (2008-present). He has been the supervisory editor of the Media Ecology Book Series published by Hampton Press (1995-2011), edited the second edition of Alfred Korzybski's *Selections from Science and Sanity* (2010), and is a partner in NeoPoiesis Press (2009-present). He maintains a blog about media, technology, language, symbols, etc., entitled *Blog Time Passing* http://lancestrate.blogspot.com> (2007-present), as well as a poetry blog http://www.myspace.com/lancestrate/blog (2007-present). He received the New York State Communication Association's John F. Wilson Fellow Award in 1998, in recognition for exceptional scholarship, leadership, and dedication to the field of communication, and Denver Mayor Wellington E. Webb proclaimed "that February 15, 2002 be known as Dr. Lance Strate Day in the City and County of Denver" in honor of the keynote address he gave for the Rocky Mountain Communication Association. Translations of his writing have appeared in French, Spanish, Italian, Hungarian, Hebrew, Chinese, and Quenya.

www.ingramcontent.com/pod-product-compliance
Lightning Source LLC
Chambersburg PA
CBHW062151080426
42734CB00010B/1646